# STUDY GUIDE and PROBLEMS WORKBOOK

to accompany

Kleinsmith/Kish
# PRINCIPLES OF CELL AND MOLECULAR BIOLOGY
*Second Edition*

Christine L. Case
*Skyline College*

Supplements Editor: Donna Campion

Key figures: J/B Woolsey Associates

Copy Editor: William Tucker


Study Guide and Problems Workbook to accompany Kleinsmith/Kish, PRINCIPLES OF CELL AND MOLECULAR BIOLOGY, Second Edition

**Copyright © 1995 HarperCollins*Publishers***

All rights reserved. Printed in the United States of America. No part of this book may be used or reproduced in any manner whatsoever without written permission of the authors or the publisher, with the following exceptions: handouts and testing materials may be copied for classroom use. For information, address HarperCollins College Publishers, 10 East 53rd Street, New York, NY 10022.

ISBN: 0-065-00405-1

95  96  97  98  9  8  7  6  5  4  3  2  1

# Table of Contents

List of Key Figures .................................................................................................... iv

Preface ....................................................................................................................... v

Chapter 1     Prologue: Cells and Their Molecules ........................................................ 1

Chapter 2     Energy and Enzymes ............................................................................... 13

Chapter 3     The Flow of Genetic Information ............................................................ 25

Chapter 4     Experimental Approaches for Studying Cells ......................................... 41

Chapter 5     Membranes and Membrane Transport ..................................................... 57

Chapter 6     The Cell Surface and Cellular Communication ...................................... 69

Chapter 7     Cytoplasmic Membranes and Intracellular Traffic .................................. 87

Chapter 8     Mitochondria and the Capturing of Energy Derived from Food ............ 101

Chapter 9     Chloroplasts and the Capturing of Energy Derived from Sunlight ........ 115

Chapter 10    The Nucleus and Transcription of Genetic Information ......................... 129

Chapter 11    The Ribosome and Translation of Genetic Information ......................... 145

Chapter 12    Cell Cycles and Cell Division ................................................................ 159

Chapter 13    The Cytoskeleton and Cell Motility ....................................................... 173

Chapter 14    Evolution of Cells and Organelles .......................................................... 189

Chapter 15    Gametes, Fertilization, and Early Development ..................................... 201

Chapter 16    Lymphocytes and the Immune Response ............................................... 215

Chapter 17    Neurons and Electrical Signaling ........................................................... 227

Chapter 18    The Cancer Cell ...................................................................................... 239

# List of Key Figures

Structure of a Typical Animal Cell
(Figure 1-2) .................................................................................................................7

Structure of a Typical Plant Cell
(Figure 1-3) .................................................................................................................7

The Lock-and-Key and Induced-Fit Models of Enzyme-Substrate Interaction
(Figure 2-4) ...............................................................................................................18

Role of Topoisomerase in DNA Unwinding
(Figure 3-14) .............................................................................................................34

Comparison of Image Formation in the Light and Transmission Electron Microscopes
(Figure 4-6) ...............................................................................................................49

Image Formation in the Scanning Electron Microscope
(Figure 4-16) .............................................................................................................49

The Fluid Mosaic Model of Membrane Organization
(Figure 5-5) ...............................................................................................................64

Mechanism by Which the $G_s$ Protein Mediates the Interaction between a Receptor
and Adenylyl Cyclase (Figure 6-17) .........................................................................80

Diagram Showing How the Endoplasmic Reticulum Divides the Cytoplasm into
Two Compartments, the Cisternal Space and the Cytosol (Figure 7-4) ...................95

Model of the Respiratory Chain in the Inner Mitochondrial Membrane
(Figure 8-30) ...........................................................................................................108

Model of Electron Transfer and ATP Formation in the Thylakoid Membrane
(Figure 9-29) ...........................................................................................................123

Summary of the Positive and Negative Control Systems on the *lac* Operon
(Figure 10-52) .........................................................................................................138

The Elongation Phase of Protein Synthesis
(Figure 11-30) .........................................................................................................152

The Role of Mitotic Cyclin and MPF in Triggering the Onset of Mitosis
(Figure 12-9) ...........................................................................................................166

The Sliding Filament Model of Muscle Contraction
(Figure 13-11) .........................................................................................................184

The Main Events That Might Have Occurred During the Evolution
of Eukaryotic Cells (Figure 14-36) .........................................................................196

Steps Involved in Sperm-Egg Fusion in Sea Urchins
(Figure 15-17) .........................................................................................................208

Evidence for the Existence of Antigen Receptors on the Surface of Lymphocytes
(Figure 16-14) .........................................................................................................222

Structure of Chemical and Electrical Synapses
(Figure 17-11) .........................................................................................................233

Malignant Transformation by RNA Viruses
(Figure 18-15) .........................................................................................................244

# Preface

Biology is a rewarding and challenging field. The **rewards** have been well-documented in career studies. In 1992, *Money* (magazine) did a survey of the 100 best jobs in America and biologist was rated first. Their criteria included salary, prestige, working conditions, work schedules, and job security. Moreover, in 1993 *Money* reported that biologist is one of the 20 jobs that will show the greatest percentage employment gains by 2005.

The **challenge** is to take an active part in your learning. One semester, Thomas came into my lab before class and took the seat in the back corner—near the exit and as far away from me as physically possible. During the semester, Thomas appeared bored during lecture and usually left lab early, rarely doing more than opening his lab manual. Alas, he didn't pass that semester and returned the following semester. This time, he took the front-row center seat. He responded to questions during lecture and worked with his classmates during lab. And, this time Thomas earned an A. The course was no easier or harder the second time and certainly he didn't learn enough during his first try to make the second time easier. What happened? Thomas took responsibility for his learning. He did not sit passively, that *is* boring. He kept his thoughts on the subject throughout class and as he learned he wanted to learn more. Now he wanted to be able to answer his own questions, not just mine.

During another semester, Andrea always came to class eager to participate. She sat on the edge of her chair through lecture and worked with her classmates in lab. and in out-of-class study sessions. Her study group always had the highest scores because of their study habits and their friendly competition.

Like Andrea and Thomas, you won't learn the material simply because you purchased the text book or this Study Guide or taped the lectures. Remember that taking notes is not just to have a reference for later use but a technique to make the learner actively participate. And tests are designed to promote studying and learning, they are not just so your professor can bestow a grade.

The questions in this study guide are to give you practice using your new information. After you have read a chapter, answer the questions to test yourself. Study the chapter again and in a couple of days try the questions again. Your goal is to get comfortable using new vocabulary and concepts.

Andrea and Thomas exemplify the students for whom this Guide is prepared and to whom it is dedicated.

I would like to thank my colleagues, Dr. Ed Wodehouse, Bonnie Okonek, Jeanne Weidner, and Howard Sundberg, for reviewing this Guide and for sharing their test questions with me. I am indebted to Don Biederman for indulging me with many patient discussions.

If you have suggestions for this Guide, please send them to me:

Christine L. Case
Skyline College
San Bruno, CA 94066-1698
case@smcccd.cc.ca.us

## *Reviewers*
Professor Doug Carmichael, Pittsburg State University
Professor Richard H. Falk, University of California, Davis

# CHAPTER 1

# PROLOGUE: CELLS AND THEIR MOLECULES

## LEARNING OBJECTIVES

Be able to
1. State the cell theory.
2. Describe the contributions made to the cell theory by Hooke, van Leeuwenhoek, Schleiden and Schwann, and Virchow.
3. List four basic functions performed by cells.
4. Identify the functions of the following: plasma membrane, endoplasmic reticulum, Golgi complex, lysosomes, peroxisomes, and vacuoles.
5. Differentiate between the following:
    a. Cell and protoplast
    b. Eukaryotic cell and prokaryotic cell
    c. Cell wall and plasma membrane
    d. Chromatin and chromosome
    e. Nucleus and nucleoid
    f. Nucleus and nucleolus
6. Identify the function of ribosomes and differentiate between prokaryotic and eukaryotic ribosomes.
7. Identify the functions of mitochondria and chloroplasts and differentiate between these two organelles.
8. Define the following terms: cytoskeleton, microtubules, actin filaments, intermediate filament, and cilia and flagella.
9. Identify factors that limit the size of a cell.
10. List the characteristics of water that make it especially suitable for a cell.
11. Define organic molecule and differentiate between sugars, fatty acids, nucleotides, and amino acids.
12. Define polysaccharide and differentiate between starch and cellulose.
13. Define the following terms: saturated fatty acid, unsaturated fatty acid, fatty acid, triacylglycerol, phospholipid, glycolipid, steroid, and terpene.
14. List the components of DNA and RNA.
15. Define N—terminus, C—terminus, peptide bond, and protein.
16. Describe the primary, secondary, tertiary, and quaternary structure of a protein.
17. Describe a virus and list the five steps of viral reproduction.

## CHAPTER OVERVIEW

I. AN OVERVIEW OF CELL ORGANIZATION, pp. 4-10

A. **The Discovery of Cells Led to the Formulation of the Cell Theory, pp. 4-5**
1. In 1665, Robert Hooke saw dead cells in plant material through a microscope.
2. He called the "little boxes" cells.
3. Antonie van Leeuwenhoek used microscopes to observe blood cells, sperm cells, and one-celled organisms.
4. In 1838, Mathias Schleiden and Theodor Schwann proposed the cell theory stating that all organisms are composed of cells and that all cells are structurally similar.

# Chapter 1

      5. In 1855, Rudolf Virchow added to the cell theory that all cells arise from preexisting cells.

**B. Cells Perform Four Basic Functions, p. 5**
1. The plasma membrane provides a selective barrier from the environment.
2. Genetic information is stored in DNA; DNA is reproduced before cell division.
3. Cells use enzymes to metabolize food into building blocks and energy.
4. Most cells have some type of motility.

**C. Membranes and Walls Create Barriers and Compartments, pp. 5-8**
1. The selectively permeable plasma membrane is the outer boundary of cells. The membrane regulates the flow of materials into and out of the cell.
2. The region outside the nuclear envelope and inside the plasma membrane is called cytoplasm.
3. In active transport, molecules are moved into and out of a cell against a concentration gradient. Materials can also be moved into and out of a cell by membrane vesicles formed from the plasma membrane.
4. The plasma membrane transmits signals from the environment to the cell in cell-to-cell communication.
5. Plant and bacterial cells have a cell wall outside the plasma membrane.
6. Plant and bacterial cells from which the cell wall is removed are called protoplasts; protoplasts are susceptible to breakage.
7. The internal membrane systems, consisting of the endoplasmic reticulum (ER) and Golgi complex, transport and package newly synthesized proteins, respectively.
8. Membrane-enclosed organelles include: mitochondria, lysosomes, peroxisomes, and vacuoles.
9. Lysosomes digest food and unneeded intracellular constituents.
10. Peroxisomes carry out oxidation reactions.
11. Vacuoles serve as storage compartments and to maintain water balance.

**D. The Nucleus and Ribosomes Function in the Flow of Genetic Information, p. 8**
1. Eukaryotic cells have a double-membrane bounded nucleus.
2. The nucleus contains the chromatin (consisting of DNA and protein) and the nucleolus.
3. The nucleolus is DNA involved in the formation of ribosomes.
4. Nuclear sap fills the spaces around the chromatin and nucleoli.
5. During cell division the chromatin condenses into chromosomes.
6. The nucleus is surrounded by the nuclear envelope which has nuclear pores to exchange materials between the nucleus and cytoplasm.
7. Prokaryotic cells do not have a membrane-bounded nucleus; the single chromosome is folded into a region called the nucleoid.
8. Protein synthesis occurs at the ribosomes; prokaryotic ribosomes are slightly smaller than eukaryotic ribosomes.
9. In eukaryotic cells, ribosomes are found in the cytoplasm and attached to the ER; prokaryotic-type ribosomes are found in mitochondria and chloroplasts.

**E. Mitochondria and Chloroplasts Play Key Roles in Metabolism, p. 9**
1. The initial metabolism of food occurs in the cytosol, the cytoplasm surrounding the organelles.
2. Further metabolism occurs in the mitochondria in eukaryotic cells.

Chapter 1

  3. A mitochondrion is enclosed by two membranes; the inner membrane is folded into cristae that project into the inner matrix of the mitochondrion.
  4. The matrix also contains DNA and ribosomes.
  5. Photosynthetic eukaryotic organisms convert sunlight into chemical energy in chloroplasts.
  6. Chloroplasts consist of two membranes surrounding the stroma; chlorophyll is embedded in thylakoid membranes in the stroma. Chloroplasts also contain DNA and ribosomes.

**F. Cytoskeletal Filaments Function in Cell Motility, p. 9**
  1. The intracellular network responsible for moving the eukaryotic cell's components is the cytoskeleton.
  2. Microtubules contain tubulin and are used to form the mitotic spindle and cilia and flagella. Microtubules have a "9 pairs + 2" arrangement in cilia and eukaryotic flagella.
  3. Smaller filaments made of actin are called actin filaments or microfilaments.
  4. Intermediate filaments provide structural support.
  5. Prokaryotic cells have thin flagella made up of a single structural protein called flagellin.

**G. Prokaryotic and Eukaryotic Cells Differ in Their Structural Complexity, pp. 9-10**
  1. Eukaryotic cells are usually larger than prokaryotic cells and contain membrane-bound organelles.

**H. Cells Are Limited in How Small or Large They Can Be, p. 10**
  1. Prokaryotic cells range in size from 1-10 μm; eukaryotic cells range from 3-30 μm.
  2. The relationships between cell volume and cell surface area controls the size of a cell. If the volume gets too large, the surface cannot provide adequate exchange of nutrients and wastes between the cell and the environment.
  3. Some cells increase their surface area with projections of the plasma membrane called microvilli.
  4. If the cell volume gets too large, the cell's contents may become too dilute; compartmentalizing metabolic functions divides the cytoplasm into smaller volumes.
  5. If the cell volume gets too large, the DNA may not be able to guide all the cell's activities; large cells may have more DNA per chromosome, more copies of chromosomes per cell, or more than one nucleus per cell.
  6. The lower limit of cell size is controlled by the number (500-1,000) of enzymes needed to function; mycoplasmas, 0.2-0.3 μm, have this minimum amount of protein.

**II. MOLECULAR COMPOSITION OF CELLS, pp. 10-27**
  1. Cellular macromolecules are composed primarily of carbon, hydrogen, oxygen, nitrogen, phosphorus, and sulfur.

**A. Water Is the Most Abundant Substance in Cells, pp. 11-13**
  1. Cells consist of 70-80% water.
  2. Water is a polar molecule in which one region has a partial negative charge and another region has a partial positive charge.
  3. Water forms hydrogen bonds between the oxygen of one molecule and the hydrogen of another molecule, giving water a high surface tension.
  4. Water has a high heat capacity because the hydrogen bonds absorb heat energy before the water boils.

Chapter 1

B.  **Water Acts as Both Chemical Reactant and Solvent, pp. 13-14**
   1. Water is formed during synthesis of macromolecules by dehydration reactions.
   2. Water molecules are broken in digestion or hydrolysis of macromolecules.
   3. Water is the primary solvent for the molecules in a cell. Horowitz and Paine showed that solutes do not behave the same in cytoplasmic water as in normal water.

C.  **Most Biological Molecules are Constructed from Sugars, Fatty Acids, Nucleotides, and Amino Acids, p. 14**
   1. Cells are composed of carbon-containing or organic molecules.
   2. In 1828, Friedrich Wöhler was the first to synthesize an organic molecule (urea).
   3. A cell's macromolecules are polymers composed of monomers of sugars, fatty acids, nucleotides, and amino acids.

D.  **Sugars Are the Building Blocks of Polysaccharides, pp. 14-15**
   1. Carbohydrates are characterized by 1 carbon:2 hydrogen:1 oxygen.
   2. Monosaccharides contain three (triose) to six carbon atoms (hexose).
   3. Glucose is an energy source and building block; glucose occurs as α—glucose and β—glucose.
   4. Ribose and deoxyribose are pentoses used in making nucleic acids.
   5. Starch (in plants) and glycogen (in animals) are polysaccharides made by joining glucose molecules in α(1→4) glycosidic linkages and α(1→6) glycosidic linkages.
   6. Glycogen is found mainly in liver and muscle cells; starch is found in chloroplasts and amyloplasts.
   7. Cellulose consists of glucose molecules joined by β(1→4) glycosidic linkages.
   8. Glycosaminoglucans are common polysaccharides formed by alternating two different sugars.

E.  **Fatty Acids are Employed in the Construction of Many Types of Lipids, pp. 15-19**
   1. Lipids are soluble in nonpolar organic solvents; they are hydrophobic.
   2. Fatty acids are the simplest lipids; they consist of long carbon chains with an acidic carboxyl (—COO$^-$) group at one end. Unsaturated fatty acids contain double bonds; saturated fatty acids contain only single bonds.
   3. Triacylglycerols consist of three fatty acids joined to glycerol. Triacylglycerols are stored in animal fat cells.
   4. Phospholipids have hydrophilic (phosphate) and hydrophobic (fatty acid) regions so they are amphipathic; phospholipids are the major components of plasma membranes.
   5. Glycolipids contain a carbohydrate group; they are found in plant membranes and nervous system membranes.
   6. Steroids are 19-carbon multiple-ring compounds. Steroids are components of some animal hormones and cell membranes.
   7. Terpenes consist of five-carbon compounds; vitamin A, coenzyme Q, and carotenoids are examples of terpenes.

F.  **Nucleotides Are the Building Blocks of Nucleic Acids, p. 19**
   1. Nucleotides consist of a five-carbon sugar (pentose), a nitrogenous base, and a phosphate group.
   2. DNA contains deoxyribose and RNA contains ribose.

3. A purine (adenine or guanine) or a pyrimidine (cytosine, thymine, or, in RNA, uracil) is attached to each pentose.
4. A phosphate group is attached to the 5′ carbon of each pentose. To form a polynucleotide, the phosphate is bonded to the 3′ carbon of another pentose.

G. **Amino Acids Are the Building Blocks of Proteins, pp. 19-20**
1. An amino acid consists of the (α—) carbon attached to an amino group (—$NH_3^+$), a carboxyl group (—$COO^-$), a hydrogen atom, and a side chain (R).
2. Proteins are made up of L-amino acids.
3. There are twenty different amino acids in proteins. Based on the R group, they can be divided into nonpolar, polar, acidic, and basic amino acids.
4. Amino acids are joined by peptide bonds in dehydration reactions to form dipeptides and polypeptides.

H. **Each Protein Has a Unique Amino Acid Sequence Known as Its Primary Structure, p. 20**
1. The linear sequence of amino acids is its primary structure.
2. In 1956, Frederick Sanger developed a technique to determine the amino acid sequence in proteins.

I. **The Secondary Structure of Protein Molecules Involves α Helices and β Sheets, pp. 20-24**
1. In the 1950s, Linus Pauling discovered that hydrogen bonding between amino acids in a polypeptide causes the formation of α helix or β sheet secondary structures.
2. The linear chain loops to form α helices and β sheets in the supersecondary structure or motif.
3. In an α helix, the —C=O of each amino acid is hydrogen bonded to the —NH group of the fourth amino acid up the helix.
4. In parallel β sheet, a folded polypeptide is held together by hydrogen bonds so the N—terminus→C—terminus are in the same direction.
5. In antiparallel β sheets, the polypeptide loops run in opposite N—terminus→C—terminus directions.
6. α helices are drawn as helices or cylinders.
7. β sheets are drawn as flat arrows with the arrowhead pointing in the direction of the C—terminus.

J. **Tertiary Structure Creates Globular Proteins That are Folded into Domains, pp. 24-26**
1. A few fibrous proteins stay in the supersecondary structure. Most proteins fold into a globular arrangement called the tertiary structure.
2. Tertiary structure is due to covalent disulfide bonds, electrostatic (ionic) bonds, hydrogen bonds, Van der Waals forces, and hydrophobic effect.

K. **Quaternary Structure Creates Proteins That Consist of Multiple Subunits, p. 26**
1. The quaternary structure consists of several polypeptide subunits bonded together.

L. **Proper Conformation Is Essential for Protein Function, pp. 26-27**
1. The final three-dimensional shape of a protein is its conformation.

## Chapter 1

      2. Certain environmental conditions can destroy the protein's conformation and cause loss of function; this is called denaturation.
      3. Christian Anginsen in the 1950s discovered that some denatured proteins spontaneously regain their conformation in the proper environment.
      4. Chaperones are proteins that are involved in the proper folding of other proteins in a cell.

### III. WHAT ARE VIRUSES? pp. 27-30
      1. Viruses only reproduce using a cell's molecular machinery.

    **A. Viruses Are Infectious Agents That Are Smaller Than Cells, p. 27**
      1. In 1892, Dmitri Iwanowsky demonstrated the presence of filterable infectious agents. These were called viruses; viruses that attack bacteria are called bacteriophages.
      2. In 1935, Wendell Stanley showed that crystallized tobacco mosaic virus can infect tobacco plants.

    **B. Viruses Consist of DNA or RNA Surrounded by a Protein Coat, p. 27**
      1. Genes of viruses are either DNA or RNA.
      2. The virus's nucleic acid is surrounded by a polyhedral or helical protein coat; enveloped viruses have a membranous envelope around the protein coat.
      3. T-even bacteriophages consist of a polyhedral head attached to a helical tail.

    **C. Viruses Reproduce Inside Living Cells, pp. 28-30**
      1. To reproduce, viruses must adsorb onto a chemical receptor on a host cell.
      2. The viral nucleic acid enters the cell by penetration.
      3. In a virulent virus, the viral nucleic acid directs synthesis of viral nucleic acid and proteins. In a lysogenic virus, viral nucleic acid becomes incorporated in the host cell's DNA.
      4. During maturation, viral nucleic acid and proteins come together to form viruses. Heinz Fraenkel-Conrat demonstrated spontaneous maturation in the 1950s.
      5. Viruses are released from the cell by lysis or budding. In budding, the virus obtains part of its envelope from the host cell's plasma membrane.

    **D. Viruses May Have Evolved from Ancient Cells, p. 30**
      1. Viruses may have evolved from a precellular form of life.
      2. Or, from a preexistent cell as a degenerate cell or cell fragment.
      3. Or, viruses may be pieces of DNA or RNA that have escaped from cells.

## ☞ KEY TERMS

| | | |
|---|---|---|
| α helix | actin filament | adenine |
| amino acid | amphipathic | β sheet |
| C—terminus | carbohydrate | cell theory |
| cell wall | cellulose | chloroplast |
| chromatin | chromosomes | cilia |
| conformation | covalent | cytoplasm |
| cytosine | cytoskeleton | cytosol |
| dehydration reaction | denaturation | deoxyribose |
| disulfide bond | DNA | electrostatic bond |
| endoplasmic reticulum | eukaryotic cell | fatty acid |

## Chapter 1

flagella
glycolipid
hydrogen bond
hydrophobic
macromolecule
mitochondrion
N—terminus
nucleoid
nucleus
peroxisome
plasma membrane
polysaccharide
protein structure
pyrimidine
RNA
steroid
thymine
uracil
virus

glucose
Golgi complex
hydrolysis reaction
intermediate filament
microtubule
monomer
nuclear envelope
nucleolus
organelle
phosphodiester bond
polar molecule
prokaryotic cell
protoplast
ribose
saturated fatty acid
sugar
triacylglycerol
vacuole

glycogen
guanine
hydrophilic
lysosome
microvilli
monosaccharide
nuclear sap
nucleotide
peptide bond
phospholipid
polymer
protein
purine
ribosomes
starch
terpene
unsaturated fatty acid
Van der Waals force

## KEY FIGURES

Label these structures in the following figures. Identify the plant and animal cells.

Actin filaments
Chromatin
Golgi complex
Microvilli
Nucleolus
Plasma membrane

Centrioles
Cytosol
Lysosome
Mitochondrion
Nucleus
Endoplasmic reticulum

Chloroplast
Ribosomes
Microtubules
Nuclear envelope
Peroxisome
Vacuole

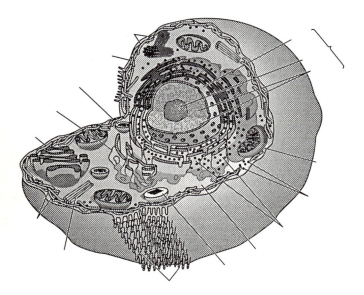
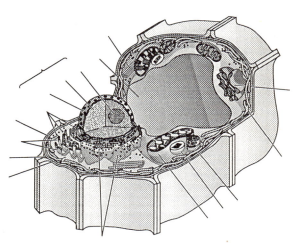

# Chapter 1

## STUDY QUESTIONS

1. Which of the following did **not** contribute to the cell theory?
   a. Hooke
   b. van Leeuwenhoek
   c. Schleiden
   d. Schwann
   e. None of the above

2. Which of the following is **not** a basic function of cells?
   a. Must have a rigid barrier from the environment
   b. Reproduce genetic material before dividing
   c. Use enzymes to metabolize food
   d. Usually have some type of motility
   e. None of the above

3. Which of the following is **not** correctly matched to its function?
   a. Peroxisome—oxidation reactions
   b. Nucleolus—ribosome synthesis
   c. Mitochondria—metabolism and production of ATP
   d. Chloroplasts—conversion of chemical energy into light energy
   e. None of the above

4. Radioisotopes are frequently used to label molecules in a cell. Assume a human cell is grown in a nutrient medium containing the radioisope $^{16}$N. After 48 hr incubation, the $^{16}$N would most likely be found in the cell's
   a. Carbohydrates.
   b. Lipids.
   c. Water.
   d. Proteins.
   e. None of the above.

5. Vinblastine is widely used for chemotherapy in treating cancer patients. Vinblastine inhibits microtubule formation. On cancer cells, this drug would prevent
   a. ER function.
   b. Cell division.
   c. Pinocytosis.
   d. Lysosome formation.
   e. Mitochondrial function.

6. The antibiotic amphotericin B which disrupts phospholipids would affect all of the following **except**
   a. Plant cells.
   b. Animal cells.
   c. Prokaryotic cells.
   d. Eukaryotic cells.
   e. None of the above.

8

7. Which one of the following is not associated with microtubules?
   a. Mitotic spindle
   b. Cilia
   c. Bacterial flagella
   d. 9+2 flagella
   e. Cytoskeleton

8. In studying a cell you find ribosomes, mitochondria, endoplasmic reticulum, and a cell wall. This cell is probably from a(n)
   a. Bacterium.
   b. Plant.
   c. Animal.
   d. Can't tell.

Use the followiing choices to answer questions 9-13.
   a. Occurs **only** in prokaryotic cells
   b. Occurs **only** in plant cells
   c. Occurs **only** in animal cells
   d. Occurs in **most** eukaryotic cells
   e. Occurs in **both** prokaryotic and eukaryotic cells

9. Cytoskeleton

10. Vacuole

11. Nuclear pore

12. 9+2 flagella

13. DNA

14. Each species of virus is specific for a particular host cell. Which step in viral replication limits the host range?
    a. Maturation
    b. Release
    c. Penetration
    d. Adsorption
    e. Replication

Use the following choices to answer questions 15-17.

a. [amino acid structure: H₃N⁺—CH(H)—COO⁻]

b. Thymine

c. [ribose/sugar ring structure with CH₂OH, OH, HO groups]

d. [phospholipid structure with two fatty acid tails, glycerol backbone, and phosphate group]

15. A building block of DNA.

16. A component of the plasma membrane.

17. A building block of a protein.

Use this figure to answer questions 18-20.

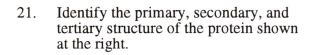

18. Is this a dehydration or hydrolysis reaction?

19. What type of organic molecules are these?

20. Identify the water molecule(s) involved in the reaction.

21. Identify the primary, secondary, and tertiary structure of the protein shown at the right.

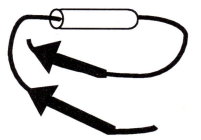

# Chapter 1

## PROBLEMS

1. In humans, brown fat, containing many mitochondria, is produced by overeating and brown fat atrophies by exposure to heat. What is the effect of overeating in a hot climate?

2. Penicillin inhibits cell wall synthesis. Why doesn't penicillin have any affect on animal cells? Why do bacterial cells die in the presence of penicillin?

3. When tested in a calorimeter, one gram of breakfast cereal has about the same number of calories as one gram of the cereal box. Aside from taste, explain why we don't eat the box.

4. Plants do not have bones. How can a terrestrial plant remain upright against gravity?

5. Gene therapy, replacing missing or defective genes, can be done by packaging the desired genes in a virus and then injecting the virus into a human. What steps in a virus life cycle make this possible?

6. Approximately 20 percent of the human population never get a cold (*Rhinovirus* infection). Suggest a reason for this.

7. Sequences of amino acids in different types of human hemoglobin are shown below. Using Figure 1-18, explain why some changes in amino acids alter function (i.e, hemoglobin S and Hammersmith hemoglobin) while other changes do not alter the function of this protein (human β chain and δ chain).

| Amino acid # | 3 | 4 | 5 | 6 | ... | 9 | ... | 40 | 41 | 42 | 43 |
|---|---|---|---|---|---|---|---|---|---|---|---|
| β chain (normal hemoglobin) | L | T | P | E | ... | S | ... | Q | R | F | E |
| Hemoglobin S (sickle-cells) | L | T | P | V | ... | S | ... | Q | R | F | E |
| Hammersmith hemoglobin (unstable) | L | T | P | E | ... | S | ... | Q | R | S | E |
| δ chain (normal hemoglobin) | L | T | P | E | ... | T | ... | Q | R | F | E |

# ANSWERS TO STUDY QUESTIONS

1. b   2. a   3. d   4. d
5. b   6. e   7. c   8. b
9. d   10. b   11. d   12. d
13. e   14. d   15. b   16. d
17. a   18. hydrolysis   19. sugars

20.

21.

# CHAPTER 2

## *Energy and Enzymes*

## ⇗ LEARNING OBJECTIVES

Be able to
1. List the four conditions needed to produce a functioning cell.
2. Define energy and thermodynamics.
3. State the first and second laws of thermodynamics.
4. Define G, ΔG, ΔG′, ΔG°′, ΔG″, and ΔG‡ and describe how they are affected by changes in total energy, temperature, and entropy.
5. Define enzyme, substrate, enzyme-substrate complex, and active site.
6. Describe the effect of each of the following on reaction rates: enzyme concentration, substrate concentration, pH, and temperature.
7. Compare and contrast the lock-and-key theory and the induced-fit model.
8. Define $V_{max}$ and $K_m$ and be able to calculate each.
9. Explain the function of coenzymes, prosthetic groups, and metal ions in enzyme activity.
10. Compare and contrast competitive inhibition and noncompetitive inhibition.
11. Explain how feedback inhibition and allosteric regulation work.
12. Define isozyme and explain how enzymes with multiple subunits work.
13. Describe the role of ATP in coupled reactions.

## ⇗ CHAPTER OVERVIEW

I. **INTRODUCTION, p. 33**
   1. Cells require chemical building blocks.
   2. Cells need energy.
   3. Cells use enzymes to speed up chemical reactions.
   4. DNA and RNA contain the information that directs a cell's activities.

II. **ENERGY FLOW IN LIVING CELLS, pp. 33-36**
   1. Energy is the capacity to do work.
   2. Thermodynamics is the study of energy transformations.

   A. **The First Law of Thermodynamics States That Energy Can Neither Be Created Nor Destroyed, pp. 33-34**
      1. The total amount of energy in the universe is constant; the kind and location of energy can be transformed or changed.

   B. **The Second Law of Thermodynamics States That the Entropy of the Universe Is Always Increasing, p. 34**
      1. Reactions proceed in the direction that increases randomness or entropy.
      2. Free energy (G, for Josiah Gibbs) is the energy that can be harnessed to do work.

   C. **Changes in Free Energy Determine the Direction in Which Reactions Proceed, pp. 34-35**
      1. Change in free energy (ΔG) = ΔE − TΔS; ΔE = total energy of a system, T = temperature in °K, ΔS = change in entropy of a system.

2. Reactions proceed in a direction that causes a decrease in free energy; $-\Delta G$.
3. Exergonic reactions have a $-\Delta G$ and occur spontaneously.
4. Endergonic reactions have a $+\Delta G$ and do not occur spontaneously.
5. Reactions with $+\Delta S$, $-T\Delta S$, or $-\Delta G$ values occur spontaneously.

D. **$\Delta G°'$ Determines the Direction in Which Reactions Proceed under Standard Conditions, pp. 35-36**
1. Standard conditions are 298°K, 1 atmosphere, and reactants and products at initial concentrations of 1.0 *M*.
2. Standard free energy change ($\Delta G°'$) is a measure of the amount of free energy released in a reaction under standard conditions.
3. $\Delta G°' = -2.303\ RT \log_{10} \frac{[\text{products}]_{eq}}{[\text{reactants}]_{eq}}$
4. $\Delta G°' = -2.303\ RT \log_{10} K'_{eq}$
5. $K'_{eq} = \frac{[\text{products}]_{eq}}{[\text{reactants}]_{eq}}$

E. **$\Delta G'$ Determines the Direction in Which Chemical Reactions Actually Proceed Inside Cells, p. 36**
1. $\Delta G'$ is used because standard conditions do not usually apply in a cell.
2. $\Delta G' = \Delta G°' + 2.303\ RT \log_{10} \frac{[\text{products}]_{eq}}{[\text{reactants}]_{eq}}$

III. **ENZYMES AND CATALYSIS, pp. 36-39**
1. Enzymes are molecules that speed up the rate of chemical reactions.

A. **Enzymes Function as Biological Catalysts, p. 37**
1. A catalyst is an agent that speeds up a chemical reaction without being consumed in the process.
2. Jon Berzelius discovered biological catalysts (enzymes).
3. In 1897, Eduard Büchner discovered that enzymes work even when extracted from a cell.
4. Enzyme names usually end in "–ase."

B. **Enzymes are Almost Always Proteins, pp. 37-38**
1. In 1926, James Sumner purified the enzyme urease. He concluded that enzymes are made of proteins; and this was confirmed by John Northrop.
2. There are six major classes of enzymes based on the types of reactions they catalyze.
3. In the 1980s, Cech and Altman discovered a group of enzymes, ribozymes, composed of RNA.

C. **Enzymes Are More Efficient, Specific, and Controllable Than Other Catalysts, pp. 38-39**
1. Catalysts are substances that increase the rate of chemical reactions, are not consumed during the reactions, work in small quantities, and do not alter the equilibrium of the reaction.
2. Enzymes speed reactions $10^8$ to $10^{11}$ times.
3. Enzymes are specific for a substance and reaction.
4. A cell can control enzyme activity.

Chapter 2

IV. **MECHANISM OF ENZYME ACTION, pp. 39-43**

  A. **Catalysts Speed Up Reactions by Lowering the Activation Energy ($\Delta G^\ddagger$), pp. 39-40**
   1. During the transition state, the free energy of the reactants is higher than that of the initial reactants.
   2. Activation energy ($\Delta G^\ddagger$) is the amount of energy required to reach the transition state.
   3. The rate of a chemical reaction is directly proportional to the number of molecules reaching the transition state.
   4. The application of heat can provide the needed $\Delta G^\ddagger$ for a reaction.
   5. Enzymes act by lowering the $\Delta G^\ddagger$ for a reaction.

  B. **Enzymes Lower Activation Energy by Reducing the Energy and/or Entropy Changes Associated with the Transition State, p. 40**
   1. $\Delta G^\ddagger = \Delta E^\ddagger - T\Delta S^\ddagger$; $E^\ddagger$ is the total energy and $S^\ddagger$ is the entropy of the transition state.
   2. Enzymes alter $\Delta E^\ddagger$ by interacting with reactants to form transition states that require less energy.
   3. By binding to reactants, enzymes increase the probability of collisions, thereby decreasing $\Delta S^\ddagger$.

  C. **An Enzyme-Substrate Complex Is Formed During Enzymatic Catalysis, pp. 40-41**
   1. In 1913, Michaelis and Menten first proposed that an enzyme binds to its substrate to form an enzyme-substrate complex.
   2. At low substrate concentrations, the rate of a reaction is proportional to the substrate concentration; as substrate concentration increases, the enzyme becomes saturated and the reaction rate does not increase.
   3. In the 1930s, Keilin verified the formation of an enzyme-substrate complex.

  D. **Enzyme Specificity Is Explained by the Shape of the Active Site, pp. 41-42**
   1. Fischer proposed that an enzyme has an active site which recognizes the three-dimensional structure of the substrate.
   2. According to the lock-and-key theory, substrate binding only occurs if the substrate (key) precisely fits the enzyme (lock).
   3. Koshland proposed the induced-fit model that states the final proper fit between enzyme and substrate only occurs when the enzyme changes shape after its initial reaction with the substrate.

  E. **The Catalytic Efficiency of the Active Site Is Based on a Combination of Factors, pp. 42-43**
   1. Enzymes bring multiple substrates together to increase the likelihood of a reaction.
   2. Enzymes can form covalent intermediates with their substrates to generate a change in substrates.
   3. An enzyme's active site may act as a donor or acceptor of protons or electrons.
   4. The presence of the enzyme may cause a strain in the substrate which increases its reactivity.

V. **ENZYME KINETICS AND REGULATION, pp. 43-55**
   1. Enzyme kinetics is the analysis of the factors which determine the rate of an enzyme-catalyzed reaction.

## Chapter 2

- **A. $V_{max}$ and $K_m$ Provide Information about an Enzyme's Catalytic Efficiency and Substrate Affinity, pp. 43-44**
    1. $V_{max}$ is the maximum velocity that a reaction can attain; $V_{max}$ varies with enzyme concentration.
    2. Turnover number is the number of substrate molecules that one enzyme can convert to product per unit time (e.g., one second).
    3. $K_m$ or Michaelis constant is the substrate concentration at which the reaction velocity is one-half maximum velocity ($V_{max}$); $K_m$ is not affected by enzyme concentration.

- **B. Values for $V_{max}$ and $K_m$ Can Be Determined Using the Michaelis-Menten Equation, pp. 44-46**
    1. $v_i = \frac{V_{max}[S]}{K_m + [S]}$; $v_i$ = initial velocity of the reaction.
    2. A Lineweaver-Burk plot of $\frac{1}{v_i}$ provides a straight line from which to determine $V_{max}$ and $K_m$.

- **C. Enzyme Activity Is Influenced by pH and Temperature, p. 46**
    1. Changes in pH can denature an enzyme.
    2. The optimum pH for most enzymes is 7.
    3. Increasing temperature up to the optimal will increase the rate of a reaction.
    4. Further increases in temperature decrease the rate of a reaction and denature the enzyme.
    5. The optimum temperature for mammalian enzymes is 37°C.

- **D. Enzymes May Require Coenzymes, Prosthetic Groups, and Metal Ions for Optimal Activity, pp. 47-48**
    1. Coenzymes are small organic molecules that bind reversibly to enzymes and participate in the catalytic process; coenzymes are derived from vitamins.
    2. Prosthetic groups are small organic molecules that bind tightly to the protein portion of an enzyme (apoprotein) and participate in the catalytic process.
    3. Some enzymes consist of a protein bound to a metal ion.

- **E. Enzyme Inhibitors Can Act Irreversibly or Reversibly, pp. 48-51**
    1. Irreversible inhibitors bind covalently and permanently to enzymes to disrupt their function.
    2. Competitive inhibitors bind reversibly to the enzyme's active site. When the competitive inhibitor is bound, the enzyme cannot react with its substrate.
    3. Noncompetitive inhibitors bind enzymes at a site other than the active site, thereby changing the shape of the active site.

- **F. Allosteric Regulation Involves Reversible Changes in Protein Conformation Induced by Allosteric Activators and Inhibitors, pp. 51-52**
    1. Inhibition of a metabolic pathway by the product of that pathway is called feedback inhibition.
    2. In 1963, Monod, Changeux, and Jacob discovered allosteric regulation.
    3. An allosteric inhibitor binds to the allosteric site of an enzyme to decrease enzyme activity; an allosteric activator binds to the allosteric site to increase enzyme activity.
    4. Gerhart and Schachman discovered that some enzymes consist of a catalytic subunit that carries out the reaction and a regulatory subunit that binds the allosteric regulator.

G. **Allosteric Enzymes Exhibit Cooperative Interactions between Subunits, pp. 52-53**
    1. Enzymes composed of multiple subunits do not usually exhibit Michaelis-Menten kinetics.
    2. Positive cooperativity occurs when the substrate causes active sites on other catalytic subunits to bind substrate.
    3. Negative cooperativity occurs when the substrate causes other active sites to have decreased affinity for substrate.

H. **Enzyme Activity Can Be Regulated by Covalent Modifications of Protein Structure, pp. 53-55**
    1. Adding and removing chemical groups and cleaving peptide bonds can alter enzyme activity.
    2. Phosphorylation and dephosphorylation are the most common modifications.
    3. Proenzymes (or zymogens) are inactive enzyme precursors.
    4. Proenzymes are made active by enzymatically cleaving peptide bonds to remove extra amino acids.

I. **Enzyme Activity Can Be Regulated by the Association and Dissociation of Subunits, p. 55**
    1. Isozymes are forms of an enzyme that consist of different subunits.
    2. Isozymes are different forms of the same enzyme.
    3. Isozymes catalyze the same reaction but differ in $K_m$ and $V_{max}$.

VI. **COUPLED REACTIONS AND THE ROLE OF ATP, pp. 55-58**
    1. Enzymes act as coupling agents that connect endergonic to exergonic reactions.

A. **ATP Plays a Central Role in Transferring Free Energy, pp. 55-56**
    1. ATP → ADP + $P_i$ is an exergonic reaction; $\Delta G^{o\prime}$ = -7.3 kcal/mol; $\Delta G'$ = -11 kcal/mol.
    2. In 1940, Lipmann postulated that energy released from metabolism of food is used to make ATP and ATP is hydrolyzed when energy is needed by the cell.

B. **The Nucleotides GTP, UTP, and CTP Are Also Involved in Transferring Free Energy, pp. 56-57**
    1. Other high-energy nucleotides that act as coupling agents are GTP, UTP, and CTP.

C. **Coenzymes Such as $NAD^+$ and FAD Play a Central Role in Coupled Oxidation-Reduction Reactions, pp. 57-58**
    1. The coenzyme $NAD^+$ is an electron acceptor in oxidation reactions. It releases electrons for reduction reactions to occur.
    2. $NADP^+$, FAD, and FMN participate in oxidation-reduction reactions.

# KEY TERMS

| | | |
|---|---|---|
| activation energy ($\Delta G^{\ddagger}$) | active site | allosteric regulation |
| allosteric site | apoprotein | ATP |
| catalyst | catalytic subunit | coenzyme |
| competitive inhibitor | dephosphorylation | endergonic |
| energy (E) | entropy (S) | enzyme |

enzyme kinetics
FAD
free energy (G)
induced-fit model
$K_m$
Michaelis-Menten
negative cooperativity
phosphorylation
prosthetic group
saturation
thermodynamics
enzyme-substrate complex
feedback inhibition
free energy change, actual ($\Delta G'$)
irreversible inhibitor
Lineweaver-Burk plot
NAD+
noncompetitive inhibitor
positive cooperativity
reduction
standard free energy change ($\Delta G^{\circ\prime}$)
transition state
exergonic
FMN
guanosine triphosphate
isozyme
lock-and-key theory
NADP+
oxidation
proenzyme
regulatory subunit
substrate
$V_{max}$

# KEY FIGURE

Identify the lock-and-key model and the induced-fit model. Label these structures in each figure: active site, enzyme-substrate complex. Where would a competitive inhibitor bind? An allosteric inhibitor?

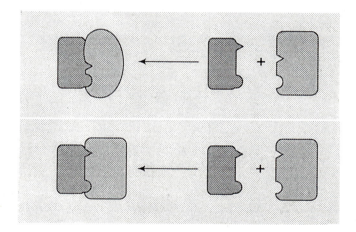

# STUDY QUESTIONS

1. Which one of the following is an enzyme?
    a. Coenzyme Q
    b. Vitamin A
    c. NAD+
    d. β-galactosidase
    e. All of the above are enzymes.

2. Which condition is **not** thermodynamically favorable for a reaction?
    a. -ΔG
    b. -TΔS
    c. +ΔS
    d. +ΔG°'
    e. All of the above

# Chapter 2

3. Which of the following reactions **cannot** occur spontaneously?
   a. Glucose-6-phosphate → fructose-6-phosphate   $\Delta G^{\circ\prime}$ = +0.4 kcal/mol
   b. ATP → ADP + $P_i$   $\Delta G^{\circ\prime}$ = -7.3 kcal/mol
   c. Pyruvate → lactose   $\Delta G^{\circ\prime}$ = -6.0 kcal/mol
   d. ADP → AMP = $P_i$   $\Delta G^{\circ\prime}$ = -7.3 kcal/mol
   e. None of the above

4. ADP + $P_i$ → ATP ($\Delta G^{\circ\prime}$ = +7.3 kcal/mol) can be made thermodynamically favorable by coupling with which of the following reactions?
   a. 3-phosphoglycerate → 2-phosphoglycerate   $\Delta G^{\circ\prime}$ = +1.0 kcal/mol
   b. Glucose 1-phosphate → glucose + $P_i$   $\Delta G^{\circ\prime}$ = -5.0 kcal/mol
   c. Phosphoenolpyruvate → pyruvate   $\Delta G^{\circ\prime}$ = -7.5 kcal/mol
   d. $CO_2$ + $H_2O$ → glucose   $\Delta G^{\circ\prime}$ = +686 kcal/mol
   e. GTP → ATP   $\Delta G^{\circ\prime}$ = 0 kcal/mol

5. The role of $NAD^+$ in the following reaction is:

   $$Lactate + NAD^+ \rightarrow pyruvate + NADH$$

   a. Enzyme
   b. Energy source
   c. Substrate
   d. Coenzyme
   e. None of the above

6. Which graph shows an enzyme at saturation?

Use the following graphs to answer questions 7-9. The graph on the left shows the optimal rate for this enzyme.

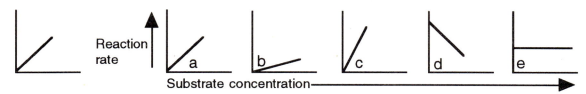

7. Which graph shows the presence of an inhibitor?

8. Which graph shows enzyme activity at a lower temperature?

9. Which graph shows the effect of decreasing the enzyme concentration by one-half?

# Chapter 2

Use the following choices to answer questions 10-11:

| Choice | Enzyme | $K_m$ | Turnover number |
|---|---|---|---|
| a. | Carbonic anhydrase | 2.6 x 10⁻² | 600,000 |
| b. Ribonuclease | | 7.9 x 10⁻³ | 790 |
| c. | Catalase | 2.5 x 10⁻² | 93,000 |
| d. | Chymotrypsin | 1.5 x 10⁻² | 0.14 |

10. Which enzyme has the highest affinity for its substrate?

11. Which enzyme is the most efficient?

12. Catalase is composed of a protein and a heme. The heme consists of a ringed protoporphyrin and an iron atom. Which of the components is a prosthetic group?
    a. Protein
    b. Heme
    c. Protoporphyrin
    d. Iron
    e. None of the above

13. The enzyme acetylcholinesterase catalyzes the breakdown of acetylcholine to acetic acid and choline. Which of the compounds shown would be the most effective competitive inhibitor?

Acetylcholine    a. Ammonium    b. Methylamine    c. Dimethylamine

d. Trimethylamine    e. Prostigmine

20

# Chapter 2

14. The enzyme catechol oxidase catalyzes the breakdown of catehol to benzoquinone that turns fruit brown. Phenol inhibits this reaction because it acts as a(n)

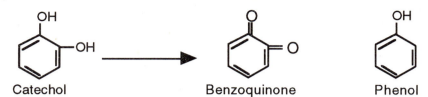

   a. Competitive inhibitor.
   b. Noncompetitive inhibitor.
   c. Allosteric inhibitor.
   d. Allosteric activator.
   e. Heavy metal

15. The enzyme threonine deaminase removes —$NH_2$ from threonine in the first reaction in the pathway shown below.

   Threonine → Intermediate 1 → Intermediate 2 → Intermediate 3 → Intermediate 4 → Isoleucine

   Isoleucine inhibits this enzyme by
   a. Competitive inhibition.
   b. Denaturing the enzyme.
   c. Feedback inhibition.
   d. Allosteric activation.
   e. Reversing the reaction.

16. In the stomach, pepsinogen is converted to the proteolytic enzyme pepsin. Pepsinogen is a(n)
   a. Isozyme.
   b. Proenzyme.
   c. Allosteric activator.
   d. Enzyme.
   e. None of the above.

17. Lactate dehydrogenase (LDH) in muscle cells consists of four M subunits and LDH in heart cells consists of four H units. These forms of LDH are called
   a. Isozymes.
   b. Proenzymes.
   c. Allosteric activators.
   d. Enzymes.
   e. None of the above.

18. Which of the following types of inhibition can be reversed by the presence of excess substrate?
   a. Noncompetitive inhibition
   b. Competitive inhibition
   c. Irreversible inhibitors
   d. Increased temperature
   e. All of the above

19. One abnormal enzyme found in cancer cells is serine kinase which phosphorylates the amino acid tyrosine instead of serine. Which of the following is the best explanation for the resultant increase in cell growth?
    a. Competitive inhibition of enzymes containing serine
    b. Allosteric inhibition of enzymes containing tyrosine
    c. Activation of enzymes containing tyrosine
    d. Reversing biochemical reactions
    e. A multienzyme complex

20. Copper is used in lakes to kill algae. The algicidal action of copper is most likely due to
    a. Competitive inhibition.
    b. Denaturing the enzyme.
    c. Allosteric inhibition.
    d. Allosteric activation.
    e. Reversing the reaction.

21. All of the following are true about enzymes **except**
    a. They can lower the $\Delta E^{\ddagger}$ for a reaction.
    b. They speed up chemical reactions.
    c. They lower the $\Delta G^{\ddagger}$ for a reaction.
    d. They decrease the $\Delta S^{\ddagger}$ for a reaction.
    e. One enzyme catalyzes many different chemical reactions.

# PROBLEMS

1. Normal liver cells contain 2 to 8% glycogen. Liver cells from patients with von Gierke's disease contain 40% glycogen. The glycogen is degraded if it is mixed with extracts from normal liver cells but not if it is mixed with von Gierke's cells. Explain the defect in von Gierke's disease.

2. Low density lipoproteins (LDL) in blood are taken up by cells where the cholesterol they carry is released. The cholesterol is a component of plasma membranes and steroid hormones. Excess cholesterol deposits in blood vessels which leads to atherosclerosis and heart attacks. Undeposited cholesterol in the blood can freely enter cells. Cholesterol inhibits the enzyme HMG CoA reductase which is involved in cholesterol synthesis. People with high cholesterol levels do not remove LDL from blood. Why does this cause high cholesterol? What would be the effect of growing the cells of a person with hypercholesteremia in a medium containing cholesterol?

3. Use the following data to determine $V_{max}$ and $K_m$ by a Lineweaver-Burk plot.

| Tube # | Substrate concentration [S] (mmol) | Reaction velocity (mmol/min) $v_i$ |
|---|---|---|
| 1 | 0.25 | 1.00 |
| 2 | 0.5 | 1.61 |
| 3 | 1.0 | 2.17 |
| 4 | 2.0 | 2.50 |
| 5 | 3.0 | 2.86 |

4. The following data are for the same enzyme as in question 3. This time an inhibitor is present. Determine the $V_{max}$ and $K_m$. What type of inhibitor is it?

| Tube # | Substrate concentration [S] (mmol) | Reaction velocity (mmol/min), $v_i$ with Inhibitor 1 | Reaction velocity (mmol/min), $v_i$ with Inhibitor 2 |
|---|---|---|---|
| 1 | 0.25 | 0.43 | 0.44 |
| 2 | 0.5 | 0.67 | 0.78 |
| 3 | 1.0 | 0.9 | 1.25 |
| 4 | 2.0 | 1.09 | 1.92 |
| 5 | 3.0 | 1.18 | 2.44 |

5. Methyl alcohol is poisonous because the enzyme alcohol dehydrogenase converts it to formaldehyde. Why could excess ethyl alcohol prevent methyl alcohol poisoning?

6. Glyphosate is used to kill unwanted plants and it does not affect humans. It inhibits the enzyme phosphoenolpyruvate (PEP) synthase. Provide an explanation for glyphosate's selectivity for plants. Recently glyphosate resistance has been genetically engineered into crop plants. Of what value is this?

7. Why would a physician look for LDH ($H_4$) to confirm a heart attack. What would the physician look for in a patient with muscular dystropy?

# ANSWERS TO STUDY QUESTIONS

| | | | | | | | |
|---|---|---|---|---|---|---|---|
| 1. | d | 2. | d | 3. | a | 4. | c |
| 5. | d | 6. | d | 7. | b | 8. | b |
| 9. | b | 10. | b | 11. | a | 12. | b |
| 13. | e | 14. | a | 15. | c | 16. | b |
| 17. | a | 18. | b | 19. | c | 20. | b |
| 21. | e | | | | | | |

# CHAPTER 3

## *The Flow of Genetic Information*

### 🏠 LEARNING OBJECTIVES

Be able to
1. Describe the contribution of each of the following in determining that DNA is the hereditary material: Mendel, Meischer, Zacharias, Levene, Griffith, Alloway, Avery, Mirsky, Chargaff, Hershey and Chase.
2. Describe the chemical and physical structure of DNA.
3. Define semiconservative replication. Identify the role of each of the following in DNA replication: leading strand, lagging strand, DNA polymerase, Okazaki fragments, DNA ligase, DNA primosome, single-stranded DNA binding proteins, topoisomerases.
4. Compare and contrast the following pairs of terms:
   a. DNA polymerase I and DNA polymerase III
   b. DNA primase and DNA polymerase
   c. DNA polymerase and reverse transcriptase
   d. DNA polymerase and RNA polymerase
   e. mRNA and tRNA
   f. Intron and exon
   g. Transposon and plasmid
   h. Gene and RFLP
5. Define gene, codon, anticodon, mutation, and transposable genetic element.
6. Differentiate between the production of bacterial mRNA and eukaryotic mRNA.
7. Identify the roles of mRNA, ribosomes, and tRNA in translation.
8. Describe how mutations occur.
9. Describe how mutations can be corrected by excision repair and mismatch repair.
10. Describe the Southern blotting and Northern blotting techniques.
11. Define the following terms: restriction enzyme, methylation, S1 nuclease, nucleic acid hybridization.
12. Outline the steps of gene cloning.
13. Define DNA synthesis and DNA sequencing.
14. Describe the polymerase chain reaction.
15. List at least four applications of genetic engineering other than cloning genes.

### 🏠 CHAPTER OVERVIEW

I. **IDENTIFYING DNA AS THE GENETIC MATERIAL, pp. 61-65**
   1. In the 1860s, Mendel concluded that hereditary information is transmitted in distinct entities, later called genes, which maintain their unique qualities over many generations.

   A. **The Scientist Who Discovered DNA Concluded That It Could Not Be Involved in the Transmission of Hereditary Information, p. 62**
      1. Meischer discovered a new chemical in the nuclei of cells. He named the chemical "nuclein," later called nucleic acid.

B. **The Idea That DNA Stores Genetic Information Was Widely Held by the Late Nineteenth Century, p. 62**
   1. In the 1880s, Zacharias discovered that chromosomes contained nucleic acids.

C. **The Idea That DNA Stores Genetic Information Was Rejected Shortly after It Was First Proposed, pp. 62-63**
   1. Levene and others discovered that nucleic acids are made of the nucleotides: A, G, C, and T.
   2. Insensitive analyses led scientists to believe that genes were made of proteins.

D. **Genetic Transformation Studies Revived the Idea That DNA Stores Hereditary Information, pp. 63-34**
   1. Griffith discovered genetic transformation when a dead S-strain of bacteria converted a live R-strain into S-strain cells.
   2. Alloway showed that a chemical extract of cells was responsible for transformation.
   3. Avery demonstrated that DNA from the S-strain cells transformed the R-stain cells, thus proving that DNA carries genetic information.

E. **The Idea That DNA Encodes Genetic Information Was Experimentally Verified by the Early 1950s, pp. 64-65**
   1. Mirsky determined that the amount of DNA in each cell of a given organism is a constant; and sperm and eggs contain half that amount of DNA.
   2. Chargaff found that DNA from different organisms had different nucleotide compositions; and DNA from cells of the same organism had similar nucleotide compositions.
   3. Hershey and Chase discovered that the virus bacteriophage T2 injects DNA into a host cell in order to replicate itself.

II. **DNA STRUCTURE AND REPLICATION, pp. 65-75**
   1. A cell duplicates its DNA before dividing.

A. **DNA Is a Double Helix, pp. 65-67**
   1. Chargaff's rule states the [A]=[T] and [G]=[C].
   2. In 1953, Watson and Crick determined the structure of DNA based on X-ray diffraction pictures taken by Franklin and Wilkins.
   3. DNA is a double helix consisting of two sugar-phosphate backbones connected by hydrogen bonds between nucleotides. One strand of DNA is $5' \to 3'$, and the other is $3' \to 5'$.
   4. Hydrogen bonding occurs between complementary nucleotides: A=T and C≡G.

B. **DNA Is Replicated by a Semiconservative Mechanism Based on Complementary Base-Pairing, pp. 67-68**
   1. Each strand serves as a template for DNA replication.
   2. DNA replication is semiconservative, that is, each new double helix consists of one old DNA and one newly synthesized DNA strand.
   3. Meselson and Stahl designed a method of incorporating $^{15}N$ into new DNA and separating new DNA from old DNA by isodensity centrifugation.

## Chapter 3

   C. **DNA Is Replicated by DNA Polymerase, pp. 69-71**
   1. Kornberg isolated DNA polymerase I, an enzyme that makes new DNA from a DNA template and the nucleoside triphosphates: dATP, dTTP, dGTP, and dCTP.
   2. The energy for synthesis comes from removing two terminal phosphate groups from the nucleoside triphosphates.
   3. DeLucia and Cairns discovered DNA polymerase II and DNA polymerase III.
   4. Temperature-sensitive mutants were used to determine that DNA polymerase III is responsible for DNA replication.

   D. **DNA Polymerases Catalyze DNA Synthesis in the $5' \to 3'$ direction, pp. 71-72**
   1. DNA polymerases synthesize DNA in the $5' \to 3'$ direction only.
   2. Okazaki discovered that the leading strand is synthesized continuously in the $5' \to 3'$ direction which the lagging strand is synthesized in short Okazaki fragments in the $5' \to 3'$ direction.
   3. The fragments are later covalently joined by DNA ligase and ATP to make a continuous $3' \to 5'$ strand.
   4. DNA polymerase III proofreads base-pairs and removes improperly base-paired nucleotides as a 3'-exonuclease.

   E. **DNA Synthesis Is Initiated Using Short RNA Primers, pp. 72-73**
   1. DNA primase synthesizes RNA primers using a DNA template.
   2. DNA polymerase III adds DNA to the primers.
   3. The RNA primers are removed by DNA polymerase I and replaced with DNA.

   F. **Unwinding a DNA Double Helix Requires Helicases, Single-Stranded DNA Binding Proteins, and Topoisomerases, pp. 73-74**
   1. DNA helicases bind to single-stranded DNA and DNA primase to form a primosome. The primosome moves with the replication fork unwinding the helix and synthesizing primers.
   2. Single-stranded DNA binding proteins maintain the single-strandedness of the lagging strand.
   3. Topoisomerases are nucleases that nick DNA to unwind it.

   G. **The Process of DNA Replication Is Complicated by Chromosome Structure, p. 74**
   1. Eukaryotic DNA is organized into chromatin fibers and chromosomes.

   H. **Retroviruses Employ Reverse Transcriptase to Synthesize DNA, pp. 74-75**
   1. Reverse transcriptase synthesizes DNA using an RNA template.
   2. Reverse transcriptase is used *in vitro* to make complementary DNA (cDNA).

III. **DNA TRANSCRIPTION AND TRANSLATION, pp. 75-90**

   A. **Experiments on *Neurospora* Led to the One Gene—One Enzyme Theory, pp. 75-76**
   1. In the early 1900s, Garrod discovered that patients with alkaptonuria lacked the enzyme needed to breakdown homogenistic acid.
   2. Using mutant cultures of *Neurospora crassa*, Beadle and Tatum found that each altered enzyme corresponded to a mutated gene leading them to conclude that one gene codes for one enzyme.

27

B. **The Base Sequence of a Gene Usually Codes for the Sequence of Amino Acids in a Polypeptide Chain, pp. 76-78**
   1. Pauling compared normal and sickle-cell hemoglobin using electrophoresis to discover that the proteins differ.
   2. Ingram used trypsin to digest the hemoglobins and separated the fragments by electrophoresis to discover only one peptide differed between normal and sickle-cell hemoglobin.
   3. A mutation resulting in changing one amino acid accounts for the difference between the two hemoglobins.
   4. Since hemoglobin is not an enzyme, the definition of a gene was modified to one gene—one polypeptide chain theory (except for genes that code for rRNA and tRNA).
   5. Yanofsky confirmed that genetic mutations cause altered polypeptides.

C. **Messenger RNA Carries Information from DNA to Newly Forming Polypeptide Chains, pp. 78-81**
   1. Brachet and Caspersson determined that cells involved in protein synthesis had large amounts of RNA in their cytoplasm.
   2. Newly synthesized proteins are associated with ribosomes indicating protein synthesis occurs at the ribosomes.
   3. Ribosomes consist of ribosomal RNA (rRNA) and protein.
   4. Jacob and Monod proposed that messenger RNA (mRNA) carries information from DNA to ribosomes.

D. **The Base Sequence of Messenger RNA Is Copied from DNA by the Enzyme RNA Polymerase, p. 81**
   1. RNA polymerase synthesizes single-stranded RNA from a DNA template using the ribonucleoside triphosphates: ATP, GTP, CTP, UTP.
   2. Marmur discovered that RNA polymerase copies only one strand of double-stranded DNA.

E. **Messenger RNA Becomes Associated with Ribosomes, pp. 81-82**
   1. Using bacteriophage T4, it was shown that mRNA and new proteins are associated with ribosomes.

F. **The Base Sequence of Messenger RNA Determines the Amino Acid Sequence of a Polypeptide Chain, p. 82**
   1. Using cell-free systems, Nirenberg showed that RNA carrying poly U directed synthesis of peptides containing phenylalanine.

G. **Genetic Information Encoded in DNA Is Expressed by a Two-Stage Process Involving Transcription and Translation, p. 83**
   1. During transcription, mRNA is synthesized from a DNA template.
   2. During translation, polypeptide chains are synthesized according to directions in the mRNA.

H. **RNA Splicing Often Accompanies the Production of Eukaryotic Messenger RNAs, p. 84**
   1. In bacteria, mRNA can be transcribed from continuous DNA stretches.
   2. In eukaryotic cells, polypeptide-encoding regions of DNA (exons) are interrupted by noncoding regions (introns).
   3. During transcription, RNA polymerase copies both introns and exons.
   4. The resulting RNA transcript undergoes splicing to remove introns.

## Chapter 3

**I.  The Genetic Code Is a Triplet Code, pp. 84-85**
1. Using frameshift mutations in bacteriophage T4, Crick and Brenner verified that each amino acid is coded for by three mRNA nucleotides.
2. There are 64 possible triplets coding for 20 amino acids.

**J.  The Coding Dictionary Was Established Using Synthetic RNA Polymers and RNA Triplets, p. 85**
1. Long, repetitive strands of RNA (e.g., UUUUUUUUUU) were shown to produce polypeptides containing one repeating amino acid (e.g., phenylalanine).
2. Nirenberg and Leder identified most of the amino acids coded by the 64 triplets; the remainder were identified by Khorana.

**K.  Of the 64 Possible Codons in Messenger RNA, 61 Code for Amino Acids, pp. 85-87**
1. mRNA triplets are called codons and are written in the $5' \rightarrow 3'$ direction.
2. Some amino acids are specified by more than one codon.

**L.  The Genetic Code Contains Special Stop and Start Signals, pp. 87-88**
1. The stop or nonsense codons, UAG, UAA, and UGA, do not code for any amino acid.
2. AUG is the start codon and codes for methionine (or N-formylmethionine in bacteria).

**M.  The Genetic Code Is Nearly Universal, p. 88**
1. The genetic code is nearly the same for all organisms.

**N.  Transfer RNAs Bring Amino Acids to the Ribosome during Protein Synthesis, pp. 88-89**
1. In 1957, Crick proposed the need for an "adapter" molecule to read mRNA and align the correct amino acid.
2. Hoagland demonstrated the existence of tRNA.
3. Amino acids are covalently bonded to tRNA molecules in the cytosol.
4. There is at least one tRNA for each amino acid.
5. Different tRNAs that bind to the same amino acid are isoaccepting tRNAs.

**O.  Each Transfer RNA Binds to a Specific Amino Acid and to Messenger RNA, p. 89**
1. Aminoacyl-tRNA synthetases bind an amino acid to the $3'$ end of a tRNA molecule to produce an aminoacyl-tRNA.
2. Chapeville and Lipmann proved that mRNA recognizes the tRNA rather than the attached amino acid.

**P.  Transfer RNAs Contain Anticodons That Recognize Codons in Messenger RNA, pp. 89-90**
1. Holley determined the first base sequence of a tRNA molecule.
2. tRNAs contain a triplet that is complementary to a codon; this triplet is called an anticodon. Anticodons are properly written in the $5' \rightarrow 3'$ direction.
3. tRNA have CCA at their $3'$ ends and contain 8-14 modified bases such as: pseudouridine, inosine, dihydrouridine, and ribothymidine.
4. X-ray crystallography by Rich and Klug showed the twisted "L" shape of tRNA.

# Chapter 3

**IV. DNA MUTATION, REPAIR, AND REARRANGEMENT, pp. 90-96**
  1. Survival of an individual requires accurate DNA replication, however, evolution is dependent on genetic change.

  **A. Spontaneous Mutations Limit the Fidelity of DNA Replication, pp. 90-91**
  1. Changes in the base sequence of a DNA molecule are called mutations.
  2. The spontaneous mutation rate is about 1 base pair/$10^9$ base pairs/cell division.
  3. Tautomers are alternate structures of nucleotide bases. For example, in the keto form, T pairs with A, in the enol form, T pairs with G.
  4. Excess of one nucleoside triphosphate and stem-and-loop structures also contribute to mutations.
  5. Depurination is the spontaneous breaking of the bond between deoxyribose and A or G and deamination is the removal of a base's amino group.

  **B. Mutations Are Also Caused by Environmental Chemicals and Radiation, pp. 92-93**
  1. Mutagens are mutation-causing agents.
  2. Base analogs resemble nucleotide bases but form abnormal base pairs.
  3. Base-modifying agents change bases by deamination, hydroxylation, and alkylation.
  4. Intercalating agents insert between bases and distort the structure.
  5. Ultraviolet radiation causes the formation of thymine dimers between adjacent thymines.
  6. Ionizing radiation damages DNA by causing breaks and highly reactive chemicals in the cytosol that can damage DNA.

  **C. Mutations Are Stabilized by Subsequent Rounds of DNA Replication, p. 93**
  1. An incorrect base-pairing can go undetected by a cell after DNA replication, see Figure 3-34.

  **D. Excision Repair Corrects Mutations That Involve Abnormal Bases, pp. 93-94**
  1. Excision repair involves the enzymatic removal and replacement of abnormal DNA bases.
  2. Enzymes involved include excinuclease, DNA glycosylases, AP endonuclease, DNA polymerase, and DNA ligase.

  **E. Mismatch Repair Corrects Mutations That Involve Noncomplementary Base Pairs, p. 94**
  1. In prokaryotes, DNA methylation occurs so the old strand can be recognized and the new (unmethylated) strand checked against it.
  2. In bacteria, A in GATC is methylated; in eukaryotes, C is methylated although its role appears to be in gene regulation.

  **F. Transposable Genetic Elements Promote the Rearrangement of DNA Base Sequences, pp. 95-96**
  1. McClintock discovered transposable genetic elements that move from one location to another.
  2. Transposable elements include insertion sequences and transposons.
  3. Transposase catalyzes the removal of a transposable element from one region of a DNA molecule and its insertion somewhere else.

## Chapter 3

**V. RECOMBINANT DNA AND GENE CLONING, pp. 96-105**

   **A. Nucleic Acid Hybridization, pp. 96-97**
1. Marmurs and Doty discovered that heating will separate the two strands of a DNA double helix (denaturation) and that the two strands will anneal (hybridize) when cooled.
2. This technique is employed in nucleic acid hybridization procedures where complementary single-stranded DNA or RNA molecules will hybridize (base-pair) with each other.
3. The temperature at which the transition from double-stranded to single-stranded DNA is halfway complete is the melting temperature ($T_m$).
4. S1 nuclease degrades single-stranded DNA.
5. Hybridization can occur between two strands that are not perfectly complementary under conditions of reduced stringency (temperature, salts, pH).

   **B. Southern and Northern Blotting Allow Hybridization To Be Carried Out with Nucleic Acids Separated by Electrophoresis, pp. 97-98**
1. A DNA probe is a labeled single-stranded DNA fragment that is used to detect its complementary sequence in DNA.
2. In Southern blotting (developed by Southern), DNA molecules separated by gel electrophoresis are transferred from the gel to a nitrocellulose filter. A DNA probe is allowed to hybridize with its complement. The presence of the probe on the filter is detected by laying the filter on X-ray film (autoradiography).
3. RNA is analyzed in the Northern blotting procedure.

   **C. *In Situ* and Colony Hybridization Allow Nucleic Acids to be Identified in Cells, p. 98**
1. Hybridization techniques used on nucleic acids in cells are called *in situ* hybridization.
2. *In situ* hybridization involves denaturing DNA, adding the appropriate probe, and visualizing radioactivity using microscopic autoradiography.
3. In colony hybridization, bacterial colonies are blotted onto a nitrocellulose filter. The cells are lysed and the DNA is denatured. A DNA probe can then be used to detect selected DNA that was in the cells.

   **D. The Discovery of Restriction Endonucleases Paved the Way for Recombinant DNA Technology, pp. 98-99**
1. Luria and Bertani discovered that bacteria have defenses against certain viruses.
2. Arber, Smith, and Nathans discovered that bacteria restrict the access of certain viruses by digesting the viral DNA with restriction endonucleases or restriction enzymes.
3. The bacterial DNA is not degraded because it is methylated.
4. Restriction enzymes cut DNA within the polynucleotide chain (exonucleases cut at the ends).
5. Each restriction enzyme recognizes a particular sequences of bases called the restriction site.
6. Restriction enzymes are named for their bacterial source, e.g., *Eco*R I was the first restriction enzyme isolated from *E. coli* bacteria.
7. Most restriction sites are palindromes:
               5′—AAGCTT—3′
               3′—TTCGAA—5′

# Chapter 3

    8.    Some restriction enzymes leave blunt ends, such as *Hae* III:
$$5'\text{—GG} \downarrow \text{CC—}3'$$
$$3'\text{—CC} \uparrow \text{GG—}5'$$
    9.    Some restriction enzymes leave sticky or cohesive ends, such as *Eco* R I:
$$5'\text{—G} \downarrow \text{AATTC—}3'$$
$$3'\text{—CTTAA} \uparrow \text{G—}5'$$

**E. Gene Cloning Techniques Permit Individual DNA Sequences To Be Produced in Large Quantities, pp. 99-104**
1. Boyer, Cohen, and Berg showed that DNA fragments with sticky ends can be joined together to form recombinant DNA.
2. The recombinant DNA can be put into bacteria and as the bacteria increase in number the recombinant DNA will be replicated. This is called DNA cloning.
3. A genomic library is made when DNA fragments from an organism's entire genome are cloned.
4. cDNA made from an mRNA template is used in a cDNA library.
5. A cloning vector is used to get DNA fragments into a bacterium or yeast for cloning.
6. Cloning vectors include plasmids, viruses, and cosmids. Shuttle vectors are capable of replicating in more than one type of cell.
7. A recombinant cloning vector must be capable of replicating within a cell. *E. coli* bacteria are commonly used hosts for cloning vectors.
8. Vectors usually have an antibiotic-resistance gene so the cells that have picked up the vector can be detected by their antibiotic resistance.
9. Clones with the gene of interest can be detected using a DNA probe.
10. An expression vector is used when the goal is to have the cell carrying recombinant DNA produce the protein product.

**F. Rapid Procedures Exist for Determining the Base Sequence of DNA Molecules, pp. 104-105**
1. Maxam and Gilbert developed the chemical cleavage method for sequencing DNA.
2. Sanger developed the chain termination method for sequencing DNA.

**G. DNA Molecules of Defined Sequence Can Be Synthesized by Chemical Procedures, p. 105**
1. Khorana was the first to create a synthetic gene. Automated DNA synthesizers are now available.

**H. The Polymerase Chain Reaction Allows Individual Gene Sequences To Be Produced in Large Quantities without Cloning, p. 105**
1. Kary Mullis developed the polymerase chain reaction (PCR) to make copies of DNA without cloning.
2. In PCR, double-stranded DNA is heated to separate the strands.
3. After cooling, a DNA primer, nucleoside triphosphates, and DNA polymerase are added.
4. This cycle can be repeated every few minutes.

# Chapter 3

## VI. PRACTICAL APPLICATIONS OF DNA TECHNOLOGY, pp. 106-110

### A. Genetic Engineering Has Facilitated the Production of Useful Proteins, p. 106
1. Proteins such as human insulin and human growth hormone have been produced by recombinant DNA techniques.

### B. Genetic Engineering in Plants Is Helping to Increase Crop Yields, p. 107
1. The Ti plasmid of *Agrobacterium tumefaciens* is used to insert recombinant DNA into plant cells.
2. Plants can be engineered for disease resistance, herbicide resistance, and improved nutritional value.

### C. DNA Fingerprinting Is a New Way of Identifying Individuals for Legal Purposes, pp. 107-108
1. DNA fingerprints are restriction enzyme fragments separated by gel electrophoresis.
2. Southern blotting is used to detect the variable number of tandem repeats which are different in each individual.

### D. Recombinant DNA Technology Is Being Used in the Diagnosis and Treatment of Genetic Diseases, pp. 108-110
1. Mutations that cause human diseases can be detected by restriction fragment length polymorphisms.
2. Gene transplantation may make it possible to treat genetic diseases.
3. A transgenic animal carries a functional gene from another organism.

## KEY TERMS

| | | |
|---|---|---|
| 3′-exonuclease | aminoacyl-tRNA | aminoacyl-tRNA synthetase |
| anticodon | AP endonuclease | autoradiography |
| base analog | base-modifying agent | blunt end |
| cDNA | cDNA library | Chargaff's rule |
| clone | cloning vector | codon |
| colony hybridization | complementary base-pairing | complementary DNA (cDNA) |
| cosmid | cytosol | deamination |
| degenerate genetic code | denaturation of DNA | depurination |
| DNA cloning | DNA fingerprinting | DNA glycosylase |
| DNA helicase | DNA ligase | DNA methylation |
| DNA polymerase | DNA primase | DNA probe |
| double helix | endonuclease | excinuclease |
| excision repair | exon | exonuclease |
| expression vector | frameshift mutation | gene |
| genomic library | *in situ* hybridization | insertion sequence |
| intercalating agent | intron | ionizing radiation |
| isoaccepting tRNA | isodensity centrifugation | lagging strand |
| leading strand | melting temperature ($T_m$) | microscopic autoradiography |
| mismatch repair | mRNA | mutagen |
| mutation | Northern blotting | nucleic acid |
| nucleic acid hybridization | Okazaki fragment | one gene—one enzyme theory |
| one gene—one polypeptide | PCR | plaque |
| plasmid | primosome | proofreading |

# Chapter 3

recombinant DNA
restriction site
RNA primer
semiconservative replication
Southern blotting
sticky end
temperature-sensitive
topoisomerases
transgenic animal
transposase
tRNA
viral cloning vectors

restriction enzyme
reverse transciptase
rRNA
shuttle vector
splicing
stop codon
template
transcription
translation
transposon
ultraviolet radiation

RFLPs
RNA polymerase
S1 nuclease
single-stranded DNA binding proteins
start codon
tautomer
Ti plasmid
transformation
transposable genetic element
triplet code
variable number tandem repeats

## ⇨ KEY FIGURE

Identify the following: DNA polymerase III, DNA helicase, DNA primase, RNA primer, DNA ligase, DNA polymerase I, Okazaki fragment, single-stranded DNA binding proteins, topoisomerases.

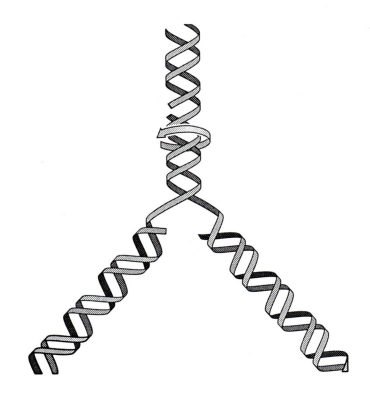

# Chapter 3

## STUDY QUESTIONS

1. Which of the following pairs is mismatched?
   a. DNA polymerase—makes a molecule of DNA from a DNA template
   b. RNA polymerase—makes a molecule of RNA from an RNA template
   c. DNA ligase—joins segments of DNA
   d. Transposase—insertion of DNA segments into DNA
   e. RNA splicing—removal of introns

2. Which of the following statements is false?
   a. DNA polymerase joins nucleotides in the 3′ → 5′ direction only.
   b. The leading strand of DNA is made continuously.
   c. The lagging strand of DNA is started by an RNA primer.
   d. DNA replication proceeds in one direction along the chromosome.
   e. Multiple replication forks are possible on a chromosome.

3. A gene is best defined as
   a. A segment of DNA.
   b. Three nucleotides that code for a polypeptide.
   c. A sequence of nucleotides in DNA that codes for a polypeptide or RNA.
   d. A sequence of nucleotides in RNA that codes for an enzyme.
   e. A sequence of nucleotides in DNA that codes for an enzyme.

For questions 4-7: Compare the paired choices for each question and answer
   a  if A is greater than B
   b  if B is greater than A
   c  if the two are nearly equal

> For example: The ease of seeing bacteria
>    A.  With a microscope
>    B.  Without a microscope          Answer:    a

4. The likelihood that ionizing radiation will cause a mutation in the following DNA sequences:
   A. ATTTCG
   B. ATCGAT

5. The likelihood of mutations after ionizing irradiation of the following DNA sequences:
   A. ATTTCG
   B. ATCGAT

6. The likelihood of a cell's progeny inheriting a change which occurred in a:
   A. nucleotide in the DNA
   B. nucleotide in the mRNA

7. The likelihood that the DNA fragment CGAATCA will hybridize with:
   A. GCTTAGT
   B. GCUUAGU

35

8. A gene contains a total of 150 phosphate molecules as part of its structure. This gene codes for a protein of approximately how many amino acids?
   a. 450
   b. 300
   c. 150
   d. 100
   e. 50

9. Which one of the following is **not** correct?
   a. RNA primers are made by RNA primase.
   b. RNA primers are removed by DNA polymerase I.
   c. Topisomerases cut DNA.
   d. Aminoacyl synthetases join amino acids to tRNA.
   e. None of the above.

10. The following steps are required for protein synthesis. What is the fourth step?
    a. Binds to mRNA
    b. Removal of introns
    c. Ribosome attaches to mRNA
    d. Transcription of DNA
    e. Translation

11. You use PCR to replicate a plasmid using $^3$H-thymidine. After three cycles, what percentage of the single strands of DNA contain a radioactive label?
    a. All
    b. 7/8
    c. 3/8
    d. 1/4
    e. None

12. A cell that cannot make tRNA
    a. Can make proteins if amino acids are provided in the growth medium.
    b. Can make proteins if mRNA is provided in the growth medium.
    c. Can't make proteins unless aminoacyl synthetase is provided in the growth medium.
    d. Can't make protein.

13. The use of an antibiotic resistance gene on a plasmid used in genetic engineering makes
    a. Transformation possible.
    b. Selection of the recombinant cell possible.
    c. The recombinant cell dangerous.
    d. The recombinant cell unable to survive.
    e. All of the above.

14. The antibiotic tetracyline binds to the 50S portion of the ribosome as shown. The effect is to
    a. Stop the ribosome from moving along the mRNA.
    b. Prevent tRNA attachment.
    c. Prevent peptide bond formation.
    d. Prevent transcription.
    e. None of the above.

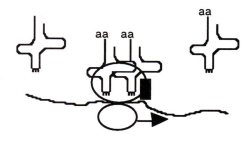

## Chapter 3

15. If you mix the following together in a test tube: DNA polymerase from *Thermus* bacteria and template DNA from a human cell, the DNA synthesized would be most similar to
    a. Human DNA.
    b. *Thermus* DNA.
    c. A mixture of human and *Thermus* DNA.
    d. Neither.
    e. Human mRNA.

16. In DNA replication, nature provides a primer in the form of
    a. DNA polymerase.
    b. Stabilizing proteins.
    c. Small strands of RNA.
    d. Nucleoside triphosphates.
    e. None of the above.

17. Which one of the following can you easily determine from the sequence of nucleotides in a bacterial gene?
    a. Primary structure
    b. Secondary structure
    c. Tertiary structure
    d. Quaternary structure
    e. All of the above

18. tRNA (anticodon: CGA) attaches to proline at 25°C and to serine at 37°C. Which of the following statements is most likely true?
    a. The organism grows best at 25°C.
    b. Translation stops at 25°C.
    c. Some proteins may not be correctly made at 25°C.
    d. Transcription will stop at 25°C.
    e. Some proteins will not be correctly made at 37°C.

19. The following steps are necessary to clone human DNA. What is the fourth step?
    a. Human DNA is inserted into a bacterial plasmid.
    b. A plasmid is inserted into *E. coli* bacteria.
    c. DNA is isolated from a human cell.
    d. Human DNA is digested with a restriction enzyme.
    e. A bacterial plasmid is digested with a restriction enzyme.

20. The vector in question 19 is
    a. A virus.
    b. A plasmid.
    c. A clone.
    d. A library.
    e. None of the above.

21. The gene for human melanin is inserted into the Ti plasmid. The plasmid is put back into *Agrobacterium tumefaciens*, what happens next?
    a. The bacterium is used to clone the gene.
    b. The bacterium inserts the gene into a plant.
    c. The bacterium causes crown gall in a plant.
    d. The bacterium inserts the gene into a human cell.
    e. The gene is transcribed and translated in the bacterium.

# Chapter 3

## 🏠 PROBLEMS

1. Using the information given, complete the sense strand of DNA and the polypeptide encoded by this strand. (Use Table 3-5.)

| base number | | 1 | 2 | 3 | 4 | 5 | 6 | 7 | 8 | 9 | 10 | 11 | 12 |
|---|---|---|---|---|---|---|---|---|---|---|---|---|---|
| DNA | (3′) | | | | | | | | | | T | T | C |
| mRNA | | | | | | | | A | C | U | | | |
| tRNA | | | | G | U | U | | | | | | | |
| amino acid | | | met- | | | | | | | | | | |

2. Explain **fully** what would be the effect of inserting a C between bases 5 and 6?

3. What is the value to a cell's survival of semiconservative replication? Of the degeneracy of the genetic code?

4. How does the experimental drug ddC (dideoxycytidine), shown below, inhibit HIV/AIDS?

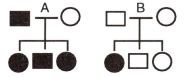

5. The neurologic disease tuberous sclerosis (TS) is a dominant trait. It can be inherited from an afflicted parent as an autosomal dominant trait (pedigree A). Explain the occurrence of TS in pedigree B.

6. Using a diagram, describe a genetic engineering experiment using the following terms: intron, exon, DNA, mRNA, cDNA, RNA polymerase, DNA ligase, restriction enzyme, splicing, recombinant DNA, translation, plasmid, reverse transcriptase.

7. After sequencing DNA, it is necessary to look at the bases to find useful restriction sites and key base sequences. How many restriction fragments will you get if you cut the following pieces of DNA with *Hae* III (GG↓CC)? How many of the pieces will hybridize completely? Can you find a start sequence? A stop sequence?

    A   5′   AATTGGCCAATAGGCCTATCCGGTAGGCC

    **B   5′   GGAAGGCCTATTGGCCAACCACATACC**

# ☞ ANSWERS TO STUDY QUESTIONS

| 1.  | b | 2.  | a | 3.  | c | 4.  | c |
| 5.  | a | 6.  | a | 7.  | c | 8.  | e |
| 9.  | a | 10. | a | 11. | b | 12. | d |
| 13. | b | 14. | b | 15. | a | 16. | c |
| 17. | a | 18. | c | 19. | b | 20. | b |
| 21. | b |     |   |     |   |     |   |

# CHAPTER 4

# *Experimental Approaches for Studying Cells*

## 👉 LEARNING OBJECTIVES

Be able to
1. Compare and contrast the following terms:
   a. *In vivo* and *in vitro*
   b. Primary cell line and transformed cell line
   c. Negative staining and positive staining
2. Define hybrid cell and describe the application of hybridomas for making monoclonal antibodies.
3. Describe the seven types of light microscopes described in this chapter, and give a use for each microscope.
4. Compare the resolution of a light microscope and an electron microscope.
5. Compare and contrast transmission electron microscopy and scanning electron microscopy.
6. Describe how intracellular structures can be located using stains, antibodies, and enzymes.
7. Compare and contrast velocity centrifugation, isodensity centrifugation, and differential centrifugation.
8. Provide an application for the use of each of the following in biology: spectrophotometry, radioisotopes, electrophoresis, and X-ray crystallography.
9. Compare and contrast agarose electrophoresis, SDS polyacrylamide electrophoresis, and two-dimensional polyacrylamide electrophoresis.
10. Describe the seven types of chromatography discussed in this chapter.

## 👉 CHAPTER OVERVIEW

   I. **GROWING CELLS IN THE LABORATORY, pp. 114-117**
   1. *In vivo* experiments are carried out while cells are within intact animals or plants.
   2. *In vitro* experiments are carried out "in glass," that is, under artificial laboratory conditions.
   3. Cell cultures are cells that are grown *in vitro*.

   A. **Animal, Plant, and Bacterial Cells Can Be Grown in the Laboratory, pp. 114-116**
   1. Harrison and Carrel grew cells *in vitro* in the early 1900s.
   2. To prepare a cell culture:
      a. Tissues must be disrupted mechanically.
      b. Animal cells are separated by proteases.
      c. $Ca^{2+}$-binding agents are used to remove $Ca^{2+}$.
      d. Desired cells can be separated with a fluorescence-activated cell sorter.
   3. Cells may be grown suspended in a liquid nutrient medium or attached to a solid surface in a nutrient medium.
   4. Animal cells form a monolayer on a solid surface.
   5. Bacteria can be grown on a nutrient medium solidified with agar.

# Chapter 4

      6. Many bacteria require only glucose, inorganic salts containing N, P, S, Mg, Na, and K, and water.
      7. Eagle defined the complex media requirements for mammalian cells; these include salts, amino acids, vitamins, and carbohydrates.
      8. Most eukaryotic cells require growth factors that are normally produced by tissues.

  **B. Primary Cell Cultures and Transformed Cell Lines Differ in Their Properties, pp. 116-117**
      1. Primary cultures grow *in vitro* for a limited period of time in culture media.
      2. Transformed cells, obtained from tumor tissue or from altered primary cultures, grow indefinitely *in vitro*.
      3. The HeLa cell line was established in 1951.

  **C. Hybrid Cells Can Be Treated by the Technique of Cell Fusion, p. 117**
      1. In 1960, Barski discovered that cells of two different types can spontaneously fuse.
      2. Harris discovered cell fusion could be induced by treating cells with mutated Sendai virus
      3. Fused cells create a heterokaryon containing two nuclei, then the nuclei fuse to create a hybrid cell.

**II. VIEWING CELLS WITH A MICROSCOPE, pp. 117-128**

  **A. The Light Microscope Can Visualize Objects as Small as 200 Nanometers in Diameter, pp. 118-119**
      1. The unaided eye cannot resolve objects smaller than 100 μm..
      2. In a light microscope, light passes through the specimen and the image is magnified by an objective lens and an ocular lens.
      3. Resolving power (*r*) is the ability to distinguish adjacent objects: $r = \frac{0.61 \lambda}{n \sin \alpha}$ where $n \sin \alpha$ = numerical aperture.
      4. Stains are used to provide contrast in specimens for microscopic examination.
      5. Prior to staining, fixatives are used to kill cells and preserve their appearance.
      6. Tissues are sliced into thin sections using a microtome.

  **B. Specialized Kinds of Light Microscopy Permit the Visualization of Living Cells, pp. 119-122**
      1. Bright field light microscopy uses visible light.
      **2. Phase contrast microscopy detects differences in refractive index and thickness, p. 119**
          a. Refractive index is a measure of the velocity with which light waves pass through a material.
          b. Interference occurs when two beams of light combine to reinforce or cancel one another.
          c. Phase contrast microscopy converts phase differences into differences in brightness.
          d. In a phase contrast microscope, refracted light (passing through the specimen) is separated from unrefracted light (by interference).
          e. Differences in phase are enhanced by a phase plate in the microscope.

3. **Polarization Microscopy Uses Polarized Light To Detect Structures That Are Birefringent, p. 120**
    a. Light whose waves travel in a single plane is polarized.
    b. Materials that rotate polarized light are birefringent.
    c. In polarization microscopy, polarized light is passed through a specimen and only light that is rotated by the specimen is visible.
4. **Interference Microscopy Detects Tiny Differences in Refractive Index, p. 120**
    a. In interference microscopy, a split beam of light is used: one passes through the specimen and the control beam passes through the medium only. When the two beams recombine, the specimen beam interferes with the control beam.
    b. The differential or Nomarski interference microscope enhances contrast by recombining a split beam of polarized light.
5. **Dark Field Microscopy Detects Scattered Light Deflected from a Specimen, p. 121**
    a. In dark field microscopy, only light that has been scattered by the specimen reaches the observer's eye.
6. **Fluoresence Microscopy Detects the Location of Specific Molecules, p. 121**
    a. A fluorescent molecule absorbs light of one wavelength and emits it at a longer wavelength.
    b. In fluorescence microscopy, cells are usually stained with a fluorescent molecule to locate a specific cellular component.
    c. A light absorbed by the fluorescent molecule is used in a fluorescence microscope.
7. **The Confocal Scanning Microscope Eliminates Blurring by Focusing the Illuminating Beam on a Single Plane within the Image, pp. 121-122**
    a. The confocal scanning microscope uses a laser beam to illuminate a single plane of a specimen at a time.
    b. A three-dimensional image can be obtained by sequentially illuminating different focal planes.

C. **The Electron Microscope Can Visualize Objects as Small as 0.2 Nanometers in Diameter, p. 122**
   1. The maximum resolution using visible light is 200 nm (0.2 μm).
   2. Electrons have a shorter wavelength than light and can achieve a maximum resolution of 0.2 nm (0.0002 μm).
   3. In an electron microscope, magnetic coils focus a beam of electrons on the specimen.

D. **The Transmission Electron Microscope Forms an Image from Electrons That Pass through the Specimen, p. 122**
   1. The TEM forms an image as electrons that are transmitted through the specimen strike a fluorescent screen or photographic plate.
   2. Areas that prevent electrons from passing through are called electron-opaque and appear as dark regions in the photographic image.
   3. Elements with high atomic numbers scatter electrons more; they are electron-opaque.
   4. Specimens are stained with heavy metals such as osmium, uranium, and lead.

- E. **Specimens Are Usually Sectioned and Stained Prior To Examination with the Transmission Electron Microscope, pp. 122-123**
    1. Specimens are fixed in metal oxides, embedded in resin, and sliced in an ultramicrotome for viewing in a TEM.
    2. TEM images are two-dimensional; stereo electron microscopy can be achieved by photographing a specimen at two different angles and superimposing the photographs with a stereo viewer.
    3. Small particles and organelles can be viewed without thin-sectioning.
    4. In negative staining, the heavy metal stains the background and not the specimen.
    5. In positive staining, the electron-opaque stain stains the specimen, leaving the background unstained.
    6. In shadow casting, a heavy metal is sprayed on one side of isolated particles or macromolecules

- F. **Freeze-Fracturing and Freeze-Etching Are Useful Techniques for Examining the Interior Organization of Cell Membranes, pp. 123-124**
    1. In freeze-fracturing, specimens are rapidly frozen and hit with a knife edge so they fracture along lines of natural weakness.
    2. When a plasma membrane is freeze-fractured, the P (protoplasmic) face and E (exoplasmic) face are revealed. The P face is the membrane half adjacent to the cytoplasm; the E face is the opposite side of the membrane.
    3. In freeze-etching, a frozen, fractured sample is exposed to a vacuum which removes the water to accentuate surface detail.
    4. In deep-etching, more prolonged exposure in the vacuum removes more water to reveal structures deeper in the specimen.

- G. **The Scanning Electron Microscope Reveals the Surface Architecture of Cells and Organelles, p. 124**
    1. Using a fine beam of electrons deflected off a specimen's outer surface, the SEM produces a three-dimensional image of the surface features.
    2. Specimens are prepared by freeze-drying and shadowed with a heavy metal.
    3. The resolution is 2-10 nm.

- H. **Scanning Probe Microscopes Reveal the Surface Features of Individual Molecules, pp. 125-128**
    1. In 1981, Binnig and Rohrer invented the scanning tunneling microscope.
    2. A probe leaks electrons over the surface of a sample about 1 nm from the sample.
    3. A computer measures the surface to produce an image.
    4. In an atomic force microscope, the scanning tip is on the surface of the object to measure surface features.

III. **USING MICROSCOPY TO LOCALIZE MOLECULES INSIDE CELLS, pp. 128-133**
    1. Cytochemical procedures are used to identify chemicals and metabolic activities by microscopy.

- A. **Staining Reactions Can Localize Specific Kinds of Molecules within Cells, pp. 128-130**
    1. Basic dyes can detect acids such as acidic phosphate groups in DNA and RNA.
    2. Acridine orange stains DNA (green fluorescence) and RNA (red fluorescence) by binding nitrogenous bases.

# Chapter 4

       3. The Feulgen reaction stains DNA.
       4. Stains for proteins usually bind amino acid side chains.
       5. Polysaccharides are stained using the periodic acid-Schiff technique.
       6. Lipid soluble dyes are used to detect lipids.

  B. **Antibodies Are Powerful Tools for Localizing Proteins and Other Antigens inside Cells, pp. 130-131**
       1. Antibodies linked to a fluorescent dye are used to detect proteins in a cell.
       2. Antibodies can be linked to peroxidase or electron-opaque metals.
       3. A specimen can be reacted first with a primary antibody then a fluorescent secondary antibody against the primary antibody.

  C. **Cytochemical Procedures Can Localize the Activity of Specific Enzymes within a Cell, pp. 131-132**
       1. Enzymes can be located within cells by techniques that assay for the enzyme's activity without removing the enzyme from the tissue.
       2. The Gomori technique is used to locate acid phosphatase. Inorganic phosphate released from the enzyme's substrate is reacted with lead nitrate. The resultant lead phosphate is electron-opaque and can be seen in electron microscopy or colored to see in light microscopy.

  D. **Microscopic Autoradiography is Employed to Locate Radioactive Molecules within a Cell, pp. 132-133**
       1. After radioactive compounds become incorporated into cells, the tissue is sectioned and coated with photographic emulsion. Locations of the radioisotope are readily visible by electron and light microscopy.
       2. The most widely used radioisotope is $^3H$ which can be incorporated into amino acids, nucleosides, and glucose to be taken up by a cell.

IV. **CENTRIFUGATION AND SUBCELLULAR FRACTIONATION, pp. 133-139**
       1. Organelles can be isolated by subcellular fractionation which usually employs centrifugation.

  A. **Differences in Size and Density Are the Main Factors That Govern the Behavior of Particles during Centrifugation, pp. 133-135**
       1. Svedberg developed a high speed centrifuge called an ultracentrifuge. The ultracentrifuge is capable of 100,000 rpm.
       2. The rate of movement (velocity) of a particle subjected to centrifugal force is given by Stokes formula. The velocity is proportional to the sedimentation coefficient ($s$) which include the radius and density of the particle and the density and viscosity of the suspending medium, and the centrifugal force ($g$).
       3. g is proportional to the rpm and radius of the rotor.
       4. A Svedberg unit (S) is a sedimentation coefficient of $10^{-13}$ sec.
       5. The Svedberg equation can be used to calculate molecular weight.

  B. **Velocity Centrifugation Separates Organelles and Molecules Based Mainly on Differences in Size, p. 135**
       1. Differential centrifugation uses a successive series of centrifugations at increasing speeds to separate organelles. The largest particles fall into a pellet after the first centrifugation; smaller particles remain in the supernatant.

2. In moving-zone centrifugation, the sample is applied at the top of a solution in a centrifuge tube. Particles of varying size will migrate down the tube at different rates during centrifugation.

C. **Isodensity Centrifugation Separates Organelles and Molecules Based on Differences in Density, pp. 135-137**
   1. A density gradient is created in the suspending medium so that particles will fall to the region in which their density is equal to that of the suspending medium.

D. **Isolating Organelles Requires That Cells First Be Broken Open in an Appropriate Medium, pp. 137-138**
   1. Cells must be broken open in an isotonic, buffered medium to maintain the integrity of the organelles. A sucrose-containing medium is most often used.
   2. Homogenizing in a glass tube with a Teflon pestle, high-speed mixing, and freezing and thawing are common means of disrupting cells.

E. **Homogenates Are Separated by Differential Centrifugation into Nuclear, Mitochondrial, Microsomal, and Cytosol Fractions, pp. 138-139**
   1. Homogenates are first centrifuged at 500-1000g to separate the nuclear fraction containing nuclei, unbroken cells, and tissue debris.
   2. Centrifuging the supernatant at 10,000g will separate the mitochondrial fraction containing mitochondria, lysosomes, and peroxisomes.
   3. Centrifuging the supernatant at ~100,000g will separate the microsomal fraction (pellet) from the cytosol fraction (supernatant).

V. **TECHNIQUES FOR STUDYING MACROMOLECULES, pp. 139-150**

   A. **Spectrophotometry Can Be Used to Detect Proteins and Nucleic Acids, pp. 140-141**
      1. The concentration of materials in solution can be determined by measuring the solution's (light) absorbance with a spectrophotometer.

   B. **Radioactive Isotopes Are A Sensitive Way of Detecting Tiny Amounts of Materials, pp. 141-142**
      1. Isotopes are forms of the same element with different atomic weight; radioactive isotopes emit radiation.
      2. $^3$H, $^{14}$C, and $^{32}$P can be incorporated into molecules by living cells. The presence of the radioisotopes is detected by autoradiography, radiation detectors, and liquid scintillation counters.
      3. In a pulse-chase experiment, cells are provided with a radioisotope for a period of time and then grown without the isotope to detect metabolic pathways involving the (radioactive) element.

   C. **Antibodies Are Sensitive Tools for Detecting Specific Proteins, p. 142**
      1. In an ELISA test, antibodies are used to measure the concentration of a protein.

   D. **The Monoclonal Antibody Technique Generates Large Quantities of Identical Antibody Molecules, pp. 142-143**
      1. In 1975, Köhler and Milstein developed a technique to make monoclonal antibodies from hybrid cells.
      2. Antibody-producing B lymphocytes are fused with tumor cells.

# Chapter 4

- E. **Dialysis and Precipitation Are Used to Separate Large Molecules from Small Molecules, pp. 143-144**
  1. Dialysis employs artificial membranes that permit the movement of water and small molecules.
  2. Some molecules can be separated from a solution by precipitation with trichloroacetic acid or perchloric acid.

- F. **Electrophoresis Separates Macromolecules Based on Differences in Size and Electric Charge, pp. 144-146**
  1. Molecules can be separated by electric charge and size in electrophoresis. During electrophoresis cations migrate toward the cathode (– pole) and anions migrate toward the anode (+ pole).
  2. **Electrophoresis of Nucleic Acids, p. 145**
     a. Polyacrylamide and agarose gels can be used to separate nucleic acids.
     b. In pulsed field gel electrophoresis, the direction of the electric field is repeatedly altered to enhance separation of DNA molecules.
     c. Gels are stained with ethidium bromide to see DNA; or developed by Southern or Northern blotting techniques.
  4. **Electrophoresis of Proteins, pp. 145-146**
     a. SDS polyacrylamide gel electrophoresis is used to separate protein molecules. The SDS detergent disrupts the secondary and tertiary structure of a protein.
     b. Protein-containing gels can be visualized by staining with Coomassie blue or autoradiography.
     c. Western blotting uses antibodies to detect specific proteins in the gel.
     d. In 1974, O'Farrell designed two-dimensional gel electrophoresis to separate >1000 proteins at one time. The solution is first electrophoretically separated by pH in isoelectric focusing.
     e. Then SDS polyacrylamide gel electrophoresis is used (at right angles) to separate proteins at each isoelectric point.

- G. **Chromatography Separates Molecules Based on Differences in Their Affinities for Two Phases, pp. 146-149**
  1. Materials are dissolved in a first (liquid or gas) phase and then passed through a bed of material (the second phase). Molecules are separated by their relative affinities for the two phases.
  2. In ion-exchange chromatography, materials are passed through a column filled with a matrix material which is capable of binding to the molecules as they pass through. The binding is due to electric charge. Molecules of differing charges emerge from the column at different rates.
  3. In adsorption chromatography, the column matrix binds the molecules by forces other than electric charge.
  4. In gel filtration chromatography, the matrix, consisting of small beads with pores, separates molecules by size. Molecules too large to enter the pores flow around the beads and come out of the column more quickly than small molecules that migrate through the beads.
  5. The matrix in affinity chromatography binds specifically to the desired molecules. For example, the substrate for an enzyme could be used to bind the enzyme. In immunoaffinity chromatography, antibodies are used in the column matrix to bind a specific protein.
  6. High-performance liquid chromatography (HPLC) uses pumps to force fluid through a special column.

7. In paper and thin-layer chromatography, a sample is spotted onto paper or silica on glass, respectively. As the solvent moves over the paper or silica, molecules in the sample move at different rates depending on their size and solubility.
8. Gas-liquid chromatography is used to separate molecules that are readily volatilized.

**H. X-Ray Diffraction Reveals the Three-Dimensional Structures of Macromolecules, pp. 149-150**
1. X-rays are diffracted differently by different atoms. Crystalline materials produce X-ray diffraction patterns that can be used to interpret the structure of biological macromolecules.

# KEY TERMS

absorbance
agar
antigen
bright field light microscope
cell fusion
column chromatography
cytosol fraction
dialysis
electron microscope
ELISA
fluorescence microscopy
freeze-etching
gel filtration chromatography
homogenate
immunoaffinity chromatography
interference
isodensity centrifugation
light microscope
microtome
monoclonal antibody
negative staining
objective lens
paper chromatography
polarized
pulse-chase experiment
resolving power
SDS polyacrylamide gel
spectrophotometer
subcellular fractionation
thin sections
transmission electron microscope
ultramicrotome
X-ray diffraction

adsorption chromatography
anion
atomic force microscope
cation
cell line
confocal scanning microscope
dark field microscopy
differential centrifugation
electron-opaque
Feulgen reaction
fluorescence-activated cell sorter
freeze-fracturing
growth factor
HPLC
*in vitro*
interference microscopy
isoelectric focusing
microscopic autoradiography
mitochondrial fraction
monolayer
Nomarski interference
ocular lens
phase contrast microscopy
positive staining
pulsed field gel electrophoresis
scanning electron microscope
sedimentation coefficient
stain
supernatant
thin-layer chromatography
two-dimensional electrophoresis
velocity centrifugation

affinity chromatography
antigen
birefringence
cell culture
chromatography
cytochemical procedures
deep-etching
E face
electrophoresis
fixative
fluorescent
gas-liquid chromatography
heterokaryon
hybrid cell
*in vivo*
ion-exchange chromatography
isotope
microsomal fraction
monoclonal antibody
moving-zone centrifugation
nuclear fraction
P face
polarization microscopy
primary culture
refractive index
scanning tunneling microscope
shadow casting
stereo electron microscopy
Svedberg unit
transformed cell
ultracentrifuge
Western blotting

# Chapter 4

## 👉 KEY FIGURES

Identify the following: the path of light or electrons, condenser lens, objective lens, ocular lens, projector lens, lamp, electron gun, and image.

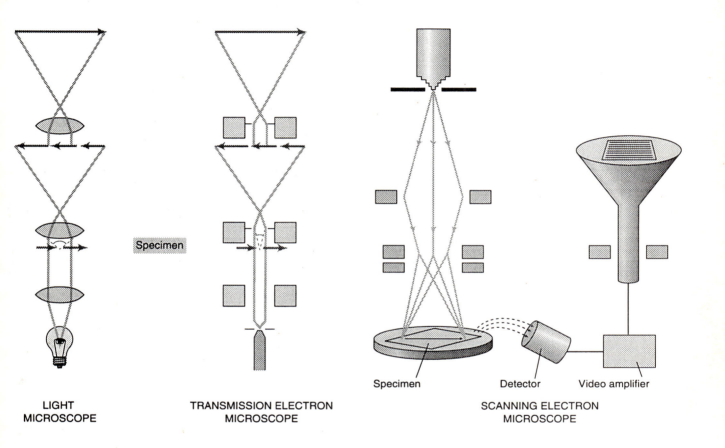

LIGHT MICROSCOPE

TRANSMISSION ELECTRON MICROSCOPE

SCANNING ELECTRON MICROSCOPE

## 👉 STUDY QUESTIONS

1. Hamster cells are grown in a nutrient medium containing the radioisotope $^{16}N$. After a 48 hour incubation, the $^{16}N$ would most likely be found in the cell's
   a. Carbohydrates.
   b. Lipids.
   c. Water.
   d. Proteins.
   e. None of the above.

49

2. Given the following information:
   | | | |
   |---|---|---|
   | White light | 500 | nm |
   | Red light | 700 | nm |
   | Blue light | 400 | nm |
   | Electrons | 0.004 | nm |

   $$\text{Resolution} = \frac{0.61 \times l}{\text{Numerical aperture}}$$

   Which of the following is needed to see a rough ER (0.25 µm) with a light microscope using an objective lens with a numerical aperture of 1?
   a. White light
   b. Red light
   c. Blue light
   d. Electrons
   e. All of them would work.

Match these microscopes with the descriptions in questions 3-7.
   a. Fluorescence microscopy
   b. Freeze-fracture followed by transmission electron microscopy
   c. Phase-contrast light microscopy
   d. Polarization microscopy
   e. Scanning electron microscopy

3. Used to see the proteins in a plasma membrane.

4. Used to see organelles in a living cell.

5. Used to see the surface structures on a 3 nm virus.

6. Used to identify sulfur crystals in a bacterial cell.

7. Used to locate rabies virus stained with labeled antibodies.

8. To extract DNA, tissue is homogenized in a high-speed mixer and centrifuged at 600g. The pellet is placed in distilled water with SDS. What is the purpose of the centrifugation?
   a. To collect the DNA in the cytosol fraction
   b. To disrupt the cells
   c. To precipitate extraneous material
   d. To precipitate DNA
   e. To collect the nuclear fraction

9. Starch is used as a source of glucose by mixing the starch with α-amylase that breaks down the starch. How would you separate the glucose molecules from the solution containing starch and α-amylase?
   a. Gel filtration chromatography
   b. Differential centrifugation
   c. SDS-polyacrylamide gel electrophoresis
   d. Electron microscopy
   e. Spectrophotometry

## Chapter 4

10. After SDS-polyacrylamide gel electrophoresis, how could you determine whether any of the bands had specific enzyme activity?
    a. Add a monoclonal antibody
    b. Add a DNA probe
    c. Add the enzyme's substrate
    d. Add a stain
    e. None of the above

11. What is the easiest method of separating out cells that have a carbohydrate-binding protein on their surface?
    a. Staining
    b. Differential centrifugation
    c. SDS-polyacrylamide gel electrophoresis
    d. Electron microscopy
    e. Affinity chromatography

12. How could you identify the human cells that possess the CD4 antigen on their surfaces and are therefore possible hosts for the AIDS virus?
    a. Use gel filtration chromatography
    b. Use monoclonal antibodies
    c. By SDS-polyacrylamide gel electrophoresis
    d. By staining
    e. By spectrophotometry

13. You are growing a human cell culture and notice after several days that one of the flasks looks different. How could you most easily determine whether your neighbor's hamster cell culture contaminated your culture?
    a. Isodensity centrifugation
    b. Light microscopy to count the chromosomes
    c. Autoradiography to locate molecules
    d. Electrophoresis
    e. Affinity chromatography

14. You are growing a cell culture that produces a chemical you want to harvest and sell. How can you separate the desired chemical from the growth medium?
    a. Differential centrifugation
    b. Column chromatography
    c. Electron microscopy
    d. X-ray crystallography
    e. Spectrophotometry

15. Which one of the following was used to determine the structure of the DNA molecule?
    a. Transmission electron microscope
    b. Scanning electron microscope
    c. Differential centrifugatioin
    d. X-ray crystallography
    e. None of the above

16. You suspect that Tay-Sachs Disease is due to excessive fats deposited in nerve cells. Which of the following is the fastest way to confirm your suspicion?
    a. Electrophoretically separate the molecules in a homogenate of the cells.
    b. Stain the cells for lipids and examine with a light microscope.
    c. Prepare thin sections for TEM examination.
    d. Centrifuge the cells.
    e. None of the above.

Match the following methods to the applications in questions 17-21.
    a. Adsorption chromatography
    b. Agarose gel electrophoresis
    c. Northern blotting
    d. Southern blotting
    e. Velocity centrifugation

17. Used to separate single-stranded and double-stranded DNA in a mixture.

18. Used to separate the various DNA fragments after treatment with a restriction enzyme.

19. Used to separate tRNA and rRNA.

20. Used to locate radioactive markers in DNA strands.

21. Used to separate chloroplasts and ribosomes.

## PROBLEMS

1. You have separated cell components by differential centrifugation and have tested each fraction for the molecules listed below. What can you conclude from the results?

|  | Nuclear fraction | Mitochondrial fraction | Microsomal fraction | Cytosol fraction |
|---|---|---|---|---|
| Protein | X | X | X | X |
| DNA | X | | | |
| RNA | | | | X |
| Succinate dehydrogenase | | X | | |
| Lysosomal acid phosphatase | X | X | X | X |
| Lactate dehydrogenase | | | | X |

2. In a pulse-chase experiment, $^{35}$S-methionine was given to a cell culture for 5 minutes and then the cells were grown for five hours in $^{35}$S-free media. Where would you expect to find the $^{35}$S-methionine immediately? After five additional hours of incubation?

3. Bacterial cells were grown in a medium consisting of oil, nitrogen and phosphorous salts. Samples were removed at 15 minute intervals. The bacterial cells were separated from the culture media and Sudan IV was added to the medium to stain the oil (red). The absorbance of each sample was then taken. How were the cells removed from the culture medium? How was the absorbance determined? What can you conclude from these data?

| Sample time (min.) | Absorbance |
|---|---|
| 0 | 1.0 |
| 15 | 0.9 |
| 30 | 0.6 |
| 45 | 0.2 |
| 60 | 0.1 |
| 75 | 0.0 |

4. The SDS polyacrylamide gel electrophoresis patterns (at the right) were obtained from the culture medium in which *E. coli* bacteria were grown.

   Lane A—Human insulin-like growth factor
   Lane B—*E. coli* culture medium after 2 hr.
   Lane C—*E. coli* culture medium after 3 hr.
   Lane D—*E. coli* culture medium after 4 hr.
   Lane E—*E. coli* culture medium after 5 hr.
   Lane F—*E. coli* culture medium after 6 hr.

   What does the gel show?

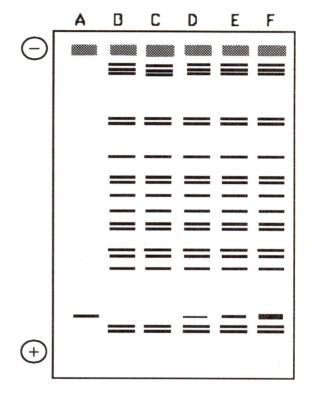

5. Two-dimensional polyacrylamide gel electrophoresis is used to separate hundreds of proteins, only a few of the proteins are shown in the gels below. Using the gel patterns (below), explain why plant A cannot fertilize itself and why it cannot be fertilized by plant B but can be fertilized by plant C.

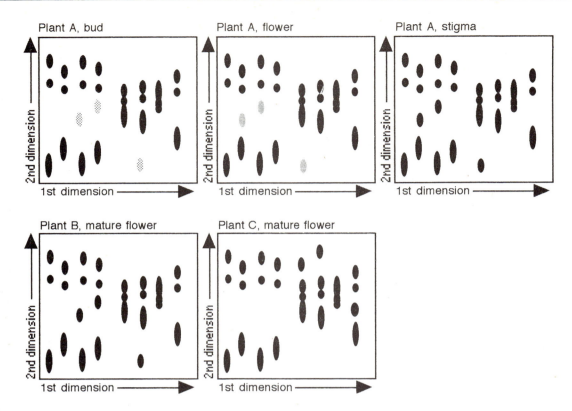

6. In transmission electron microscopy, thin sections of a specimen are examined and must be assembled to reveal the shape of the entire organism. Determine the shape of the whole organism from the thin sections shown at the right. The sections were kept in order as they were made.

7. DNA was extracted from amber and the polymerase chain reaction was used to increase the concentration of the DNA. The DNA primer used was from *Bacillus* bacteria. What can you conclude from the agarose gel pattern shown at the right? (Lane 1 is before PCR, Lane 2 is after PCR.)

# ANSWERS TO STUDY QUESTIONS

| 1. d  | 2. c  | 3. b  | 4. c  |
|-------|-------|-------|-------|
| 5. e  | 6. d  | 7. a  | 8. e  |
| 9. a  | 10. c | 11. e | 12. b |
| 13. b | 14. b | 15. d | 16. b |
| 17. a | 18. b | 19. c | 20. d |
| 21. e |       |       |       |

# CHAPTER 5

# *Membranes and Membrane Transport*

## LEARNING OBJECTIVES

Be able to
1. Describe the contributions made by the following scientists toward discovering membrane structure: Nageli, Overton, Langmuir, Gorter and Grendel, Davson and Danielli, Robertson, Schmitt and Finean, Singer and Nicolson, McConnell and Griffith, and Kornberg and McConnell.
2. Diagram a membrane showing the location of the phospholipid bilayer, integral proteins, peripheral proteins, and lipid-anchored proteins. Indicate lateral and transverse diffusion in the leaflets.
3. Compare and contrast the following: simple diffusion, facilitated diffusion, and active transport.
4. Define osmosis and describe the effects on a cell of hypotonic, hypertonic, and isotonic environments.
5. Differentiate the following pairs of terms:
   a. Anion-exchanger and ion channel
   b. Carrier protein and transmembrane channel
   c. Mobile carrier and transmembrane channel
   d. Ionophore and carrier protein
   e. Uniport and symport
   f. Symport and antiport
6. List the sources of energy for active transport.
7. Describe the action of $Na^+$–$K^+$ ATPase.
8. Describe the phosphotransferase pathway.
9. Describe the purple membrane system of *Halobacterium*.
10. Describe the $Na^+$–linked and $H^+$–linked symports.

## CHAPTER OVERVIEW

  I.  **MODELS OF MEMBRANE ARCHITECTURE, pp. 156-161**

   A. **Cellular Membranes Contain a Lipid Bilayer, pp. 156-157**
      1. Nageli *et al.* proposed the existence of a plasma membrane in the late 1800s.
      2. Overton added that the membrane is constructed of lipid molecules.
      3. Langmuir proposed the membrane is made of phospholipids which have a hydrophilic and a hydrophobic region.
      4. In 1925, Gorter and Grendel added that the membrane is a lipid bilayer with the hydrophilic groups exposed to the environment and the hydrophobic groups in the interior.

   B. **The Davson-Danielli Model Was the First Detailed Representation of Membrane Organization, p. 157**
      1. The Davson-Danielli model proposed that the lipid bilayer is covered by layers of protein and has protein-lined pores.

## Chapter 5

C. **Membranes Exhibit a Trilaminar Appearance in Electron Micrographs, pp. 157-158**
1. Membranes are 7-8 nm thick and stain as two dark lines separated by a lightly stained central zone.

D. **The Davson-Danielli Model Failed to Explain Many Aspects of Membrane Behavior, pp. 158-160**
1. The lipid bilayer is 5 nm thick; most membrane proteins are globular and must protrude from the membrane surface.
2. Many membrane proteins are hydrophobic and are bound to the hydrophobic region of the lipid bilayer. These proteins go through the lipid layer.
3. The lipid bilayer is interrupted by protein molecules.
4. 75% of the membrane surface may consist of phospholipids.
5. The hydrophilic regions of the membrane molecules are exposed to the aqueous environment.
6. Membranes are fluid structures in which the lipid and protein components can move.
7. Membranes vary in their protein/lipid ratio.

E. **The Fluid Mosaic Model Is a More Accurate Representation of Membrane Architecture, pp. 160-161**
1. In 1972, Singer and Nicolson proposed the fluid mosaic model.
2. Integral proteins are embedded in the hydrophobic regions of the lipid bilayer.
3. Peripheral proteins are bound to the membrane surface.
4. Lipid-anchored proteins are outside the lipid bilayer but are covalently bound to lipid molecules.

II. **MEMBRANE LIPIDS, pp. 161-167**

A. **Phospholipids, Glycolipids, and Steroids Are the Predominant Membrane Lipids, p. 161**
1. Phospholipids are the most prevalent lipids in membranes; sphingomyelin is a phospholipid in animal plasma membranes.
2. Glycolipids are abundant in myelin and chloroplast membranes.
3. Cholesterol is a steroid found in animal cell membranes.
4. Fatty acid chains may be saturated or unsaturated.
5. Diet and physiological conditions influence the fatty acid composition of membranes.

B. **Membrane Lipids Spontaneously Form Bilayers Because They Are Amphipathic, pp. 161-162**
1. Amphipathic substances have both hydrophilic and hydrophobic properties.
2. Amphipathic membrane lipids form bilayers which circularize as liposomes.

C. **The Lipid Bilayer Is Fluid, pp. 162-165**
1. Electron spin resonance spectroscopy showed that molecules in the membrane are fluid and not rigid.
2. Membrane lipids quickly move laterally through the cell membrane.
3. Membrane lipids flip-flop between leaflets at a slow rate.
4. The transition temperature ($T_m$) is the point at which gel-to-fluid melting occurs. The $T_m$ is lower than the normal cell's temperature.

## Chapter 5

**D. Membrane Fluidity Can Be Altered by Changes in Lipid Composition, p. 165**
1. Long fatty acid chains decrease fluidity.
2. Saturated fatty acids decrease fluidity.
3. Steroids decrease fluidity.

**E. Lipids Are Arranged Asymmetrically across the Bilayer, pp. 165-167**
1. Ghosts are empty red blood cells.
2. The outer leaflet contains saturated fatty acids, phosphatidylcholine, sphingomyelin, and glycolipids with the carbohydrate groups extending away from the cell.
3. The inner leaflet contains phosphatidylserine and phosphatidylethanolamine.

## III. MEMBRANE PROTEINS, pp. 167-175

**A. Radioactive Labeling Procedures Permit the Orientation of Membrane Proteins To Be Studied, pp. 167-168**
1. Membrane proteins are extracted from isolated membranes by SDS and separated by SDS-polyacrylamide gel electrophoresis.
2. Lactoperoxidase labels proteins with $^{125}I$ on the outer surface; or in a hypotonic solution, on the inner surface.
3. Galactose oxidase labels carbohydrates with $^{3}H$-borohydride.

**B. Membranes Contain Integral, Peripheral, and Lipid-Anchored Proteins, pp. 168-169**
1. Integral proteins are attached by hydrophobic amino acids; they span the lipid bilayer.
2. A computer-generated hydropathy plot is used to determine the α-helices containing hydrophobic amino acid clusters.
3. Polypeptide chains of peripheral proteins are entirely outside the lipid bilayer; they are bound to the hydrophilic groups of the lipids or to other proteins.
4. Lipid-anchored proteins protrude from either side of the membrane and are attached to a lipid side chain that is inserted into the membrane bilayer.

**C. Some Membrane Proteins Are Free To Move within the Lipid Bilayer, pp. 169-170**
1. Frye and Edidin demonstrated that membrane proteins diffuse through the lipid bilayer.
2. When membrane proteins cluster at one end, this is called capping.

**D. The Mobility of Membrane Proteins is Variable, pp. 170-171**
1. Fluorescence photobleaching recovery is used to determine the diffusion rate of proteins in the membrane.
2. Some proteins are restricted to separate membrane domains.

**E. Membrane Proteins Are Oriented Asymmetrically—The Red Cell Membrane as a Model, pp. 171-174**
1. SDS-polyacrylamide gels reveal few membrane proteins.
2. **The Integral Proteins of the Red Cell Membrane—Glycophorin and Band 3 Protein, pp. 172-173**
   a. Glycophorin and band 3 are glycoproteins with the carbohydrate groups facing to the exterior.

3. **The Peripheral Proteins of the Red Cell Membrane—Spectrin, Ankyrin, Band 4.1, and Actin, p. 174**
   a. Peripheral proteins are associated with the inner membrane surface.
   b. Spectrin, actin, and band 4.1 form a supportive framework for the membrane; band 4.1 is linked to glycophorin and ankyrin. Ankyrin is linked to band 3.

F. **Membrane Proteins Can Consist of Different Kinds of Subunits—The Bacterial Photosynthetic Reaction Center Complex, pp. 174-175**
   1. The photosynthetic reaction center complex of *Rhodopseudomonas viridis* consists of L, M, and H integral polypeptides.
   2. Cytochrome is a peripheral protein extending to the exterior.

IV. **TRANSPORT ACROSS CELLULAR MEMBRANES, pp. 175-192**

   A. **Diffusion Is the Net Movement of a Substance Down Its Concentration Gradient, pp. 175-176**
      1. Simple diffusion is the movement of solute molecules from a region of higher concentration (higher free energy) to a region of lower concentration.
      2. Equilibrium is the point where no further net movement occurs.

   B. **Osmosis Is Caused by the Diffusion of Water Molecules through a Semipermeable Membrane, pp. 176-177**
      1. Water diffuses from areas of low solute concentration (high water concentration) to areas of high solute concentration (low water concentration).
      2. Diffusion of water through a semipermeable membrane is called osmosis.
      3. Hypertonic solutions have a higher solute concentration than inside the cell; a hypertonic environment causes a cell to shrink.
      4. Hypotonic solutions have a lower solute concentration than inside the cell; a hypotonic environment causes a cell to swell.
      5. Isotonic solutions have the same solute concentration as inside the cell (e.g., 0.25M sucrose; 0.14M NaCl).

   C. **Simple Diffusion of Solute Molecules Is Influenced by their Lipid Solubility, Size, and Electric Charge, pp. 177-178**
      1. Using plant cells, Overton discovered that membranes are selectively permeable; in a hypertonic environment, plant cells plasmolyze.
      2. Lipid soluble materials readily diffuse across a membrane.
      3. Lipid solubility is measured by the partition coefficient (ratio of oil solubility:water solubility).
      4. Collander discovered that large molecules diffuse into cells more slowly.
      5. Membrane channels are 0.5-1.0 nm in diameter.
      6. Charged molecules do not readily diffuse across a membrane; they tend to be coated with water which decreases their lipid solubility and increases their size.

   D. **Membrane Transport Proteins Have Evolved to Aid the Transport Process, pp. 178-179**
      1. Transport of solutes by membrane transport proteins is faster than simple diffusion.
      2. The rate of transport can be described by the Michaelis-Menten equation.
      3. Membrane transport proteins are specific for a solute.
      4. Membrane transport is subject to enzyme inhibitors.

## Chapter 5

- **E. Membrane Transport Proteins Are Identified by Affinity Labeling and Membrane Reconstitution Techniques, pp. 180**
    1. A uniport is a transport protein that moves a single substance from one side of a membrane to the other, a symport protein moves two substances in the same direction, and an antiport protein moves two substances in opposite directions.
    2. In active transport, transport proteins requires energy.
    3. In facilitated diffusion, transport proteins do not require energy.

- **F. How to Determine the Energy Requirements for the Transport of Uncharged Molecules, pp. 180-181**
    1. $\Delta G' = 2.303\ RT \log_{10} \frac{[C]_{in}}{[C]_{out}}$
    2. When $\Delta G'$ is positive, transport can be coupled to hydrolysis of ATP.

- **G. How To Determine the Energy Requirements for Ion Transport, pp. 181-182**
    1. $\Delta G' = 2.303\ RT \log_{10} \frac{[C]_{in}}{[C]_{out}} + zF\Delta\psi$
    2. Concentration $\frac{[C]_{in}}{[C]_{out}}$ + charge gradient ($zF\Delta\psi$) is the electrochemical gradient.
    3. Transport that uses energy going into a cell ($+\Delta G'$) will release energy going out of the cell ($-\Delta G'$).

- **H. Facilitated Diffusion Is the Assisted Movement of a Substance Down Its Electrochemical Gradient, pp. 182-183**
    1. Ionophores are molecules which increase the diffusion rate of ions across membranes.
    2. A mobile carrier binds to an ion and diffuses across the lipid bilayer to release the ion on the other side of the membrane.
    3. Transmembrane channels allow ions to enter one side and diffuse to the other side.
    4. Valinomycin transports $K^+$; gramicidin is selective for $H^+ > NH_4^+ > K^+ > Na^+ > Li^+$.

- **I. The Glucose Transporter Mediates the Facilitated Diffusion of Glucose, pp. 183-184**
    1. Glucose transporter is a transmembrane protein.
    2. Glucose transporter protein is a carrier protein; that is, it changes shape as it carries a glucose molecule.

- **J. Band 3 Protein Is an Anion-Exchanger That Mediates the Facilitated Diffusion of Bicarbonate and Chloride Ions, p. 184**
    1. Carbonic anyhdrase converts $CO_2 \rightarrow HCO_3^- + H^+$.
    2. Band 3 protein functions as an anion-exchanger by coupling the movement of $HCO_3^-$ one direction with $Cl^-$ in the other direction.

- **K. Ion Channels Mediate the Facilitated Diffusion of Small Ions, pp. 184-186**
    1. Some protein channels are large and nonspecific.
    2. Ion channels are selective for ions.
    3. Ion channels have gates that open and close.

L. **Active Transport Involves Membrane Carriers That Move Substances against an Electrochemical Gradient, p. 186**
  1. Active transport can move substances against (up) an electrochemical gradient.

M. **ATP Can Provide Energy for Active Transport—The $Na^+$–$K^+$ Pump as a Model, pp. 187-188**
  1. Hodgkin and Keynes showed that transport of $Na^+$ out of a cell requires the presence of $K^+$.
  2. In 1957, Skou discovered the carrier protein, $Na^+$–$K^+$ ATPase.
  3. The enzyme's $Na^+$ and ATP binding sites face the inner surface of the membrane and the $K^+$ binding site faces the outer surface.
  4. 3 $Na^+$ bind and trigger phosphorylation and the enzyme moves to release the $Na^+$ outside the cell.
  5. 2 $K^+$ bind and trigger dephosphorylation and the enzyme moves to release the $K^+$ inside the cell.

N. **Phosphoenolpyruvate Can Provide Energy for Active Transport—The Phosphotransferase System as a Model, pp. 188-189**
  1. Roseman identified HPr, a small protein involved in the transfer of phosphate from phosphoenolpyruvate to sugar molecules, called the phosphotransferase pathway.
  2. The pathway requires enzymes I, II, and III in addition to HPr and chemically alters molecules that are transported.
  3. In the cytoplasm: phosphoenolpyruvate + enzyme I $\rightarrow$ enzyme I-$PO_4^-$.
  4. Enzyme I-$PO_4^-$ + HPr $\rightarrow$ HPr-$PO_4^-$.
  5. HPr-$PO_4^-$ + enzyme III$^{glu}$ $\rightarrow$ enzyme III$^{glu}$-$PO_4^-$.
  6. Enzyme III$^{glu}$-$PO_4^-$ + enzyme II$^{glu}$ $\rightarrow$ enzyme II$^{glu}$-$PO_4^-$ (at the plasma membrane).
  7. Enzyme II$^{glu}$-$PO_4^-$ + glucose (outside the cell) $\rightarrow$ glucose-$PO_4^-$ (into the cytosol).

O. **Light Can Provide Energy for Active Transport—Bacteriorhodopsin as a Model, pp. 189-190**
  1. Purple membranes of *Halobacterium* contain bacteriorhodopsin.
  2. The retinal portion absorbs light causing a conformational change in the bacteriorhodopsin.
  3. The bacteriorhodopsin transports 2 $H^+$ out of the cell. (The proton gradient can be used to generate ATP).

P. **Electron Transfer Reactions Can Provide Energy for Active Transport, p. 190**
  1. Energy for active transport can be obtained from oxidation–reduction reactions.

Q. **Ion Gradients Can Provide Energy for Active Transport—$Na^+$–Linked Transport of Sugars and Amino Acids as a Model, pp. 190-191**
  1. Christensen showed that the rate of amino acid uptake is directly related to the extracellular concentration of $Na^+$.

2. The carrier molecule has two binding sites, one for $Na^+$ and the other for the sugar or amino acid being transported. The energy is provided by the $Na^+$–$K^+$ pump.
3. The plasma membrane of intestinal cells has a $Na^+$-dependent glucose symport at the apical end; and a facilitated diffusion transporter at the basolateral surface.

R. **Proton Gradients Can Provide Energy for Active Transport—$H^+$–Linked Transport of Lactose as a Model, pp. 191-192**
1. West and Mitchell discovered that lactose uptake is accompanied by an increase in pH of the external medium.
2. Kaback confirmed that the lactose carrier transports a $H^+$ along with a lactose.

S. **Membranes Also Move Materials by the Budding and Fusion of Membrane Vesicles, p. 192**
1. In endocytosis, small regions of the plasma membrane fold inward and bud off as tiny vesicles to bring materials into the cell.
2. In exocytosis, membrane vesicles fuse with the plasma membrane to discharge materials from the cell.

# ☞ KEY TERMS

| | | |
|---|---|---|
| active transport | amphipathic | anion-exchanger |
| ankyrin | antiport | apical |
| bacteriorhodopsin | band 4.1 protein | basolateral |
| carrier protein | Davson-Danielli model | equilibrium |
| facilitated diffusion | flip-flop | fluid-mosaic model |
| ghost | glycolipid | glycoprotein |
| hydropathy plot | hydrophilic | hydrophobic |
| hypertonic | hypotonic | integral protein |
| ion channel | ionophore | isotonic |
| lateral diffusion | leaflet | lipid bilayer |
| lipid-anchored protein | liposome | membrane potential |
| mobile carrier | myelin | nonpolar |
| osmosis | partition coefficient | peripheral protein |
| phospholipase | phospholipid | phosphotransferase pathway |
| photosynthetic reaction center complex | plasma membrane | plasmolysis |
| polar | purple membrane | retinal |
| selectively permeable | simple diffusion | solute |
| spectrin | steroid | symport |
| transition temperature | transmembrane channel | uniport |

# Chapter 5

## 👉 KEY FIGURE

Identify the following: phospholipid bilayer, hydrophobic region, hydrophilic region, integral protein, peripheral protein, lipid-anchored protein, lateral diffusion, and transverse diffusion. Show the primary locations of phosphatidylcholine, sphingomyelin, phosphatidylethanolamine, phosphatidylserine, and glycolipid.

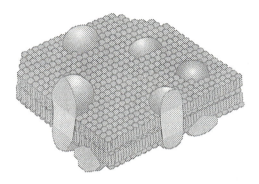

## 👉 STUDY QUESTIONS

1. Assume you are given a culture of *E. coli* bacteria that have been growing at 25°C. You incubate the culture at 37°C. What change will the bacteria make in their plasma membranes?
    a. Increase saturated fatty acids
    b. Decrease steroids
    c. Decrease length of the fatty acid chains
    d. Increase glycolipids
    e. None of the above

2. You have a preparation of ghosts and need to collect only the right-side out ghosts. You decide you can trap the right-side out ghosts in an affinity column packed with
    a. SDS.
    b. A carbohydrate-binding molecule.
    c. A protein-binding antibody.
    d. A serine-binding molecule.
    e. None of the above

3. You stain membrane proteins with a fluorescent-labeled antibody and notice the membrane is evenly stained. After a few hours, all of the fluorescence is at one end of a cell. You can conclude that
    a. Lipids flip-flop in the membrane.
    b. Proteins move laterally in the membrane.
    c. The fluorescent dye was bleached.
    d. Proteins act as transporters.
    e. All of the above

# Chapter 5

4. The fact that a red blood cell lyses when placed in an isotonic urea solution indicates that
   a. Salts leave the cell.
   b. Sugar enters the cell.
   c. Urea enters the cell.
   d. Water leaves the cell.
   e. Can't tell

5. The antibiotic amphotericin B forms channels by complexing with cholesterol. Which of the following would be killed by amphotericin B?
   a. Plant cells
   b. Animal cells
   c. Bacterial cells
   d. All of the above
   e. None of the above

6. Some bacteria are killed by detergents. In the presence of a detergent
   a. Cell contents would leak out.
   b. The cell membrane would become rigid.
   c. Water would leave the cell causing plasmolysis.
   d. Cell membranes would not be made correctly.
   e. None of the above

For questions 7-11: Compare the paired choices for each question and answer
   a  if A is greater than B
   b  if B is greater than A
   c  if the two are nearly equal

> For example: The ease of seeing bacteria
>    A.  With a microscope
>    B.  Without a microscope         Answer:    a

7. The amount of energy required for
   A. Facilitated diffusion
   B. Active transport

8. The need for a transport protein for
   A. Facilitated diffusion
   B. Active transport

9. The amount of energy required for
   A. Osmosis
   B. Transmembrane channels

10. The amount of energy required for glucose to enter the cell as
    A. Glucose
    B. Glucose-phosphate

11. The rate of diffusion into a cell of
    A. Urea
    B. $SO_4^{2-}$

12. Which of the following is **not** true?
    a. Phospholipids have hydrophilic and hydrophobic regions
    b. Proteins form continuous layers in the plasma membrane
    c. Membrane pores consist of transmembrane proteins
    d. Animal cells without cholesterol will lyse
    e. Glycolipids are used in cell recognition.

Use the following choices for questions 13-14.
    a. Antiport
    b. Symport
    c. Uniport

13. The transport protein that brings a glucose and a proton into a cell.

14. The transport protein that brings Hg+ into a bacterial cell to detoxify the mercury.

15. The bumps that are visible in freeze-fracture membrane preparations for electron micrographs are
    a. Carbohydrates.
    b. Cholesterol molecules.
    Peripheral c. ~~Extrinsic~~ proteins.
    d. Integral proteins.
    e. Phospholipids.

16. Using a light microscope, a student drew a plasma membrane as shown to the right. What is the most likely explanation for this drawing?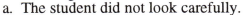
    a. The student did not look carefully.
    b. The microscope did not have high enough magnification.
    c. The student was not looking at a plasma membrane.
    d. This is an accurate depiction of a plasma membrane.
    e. None of the above

17. The poisonous heavy metal $AsO_4^{3-}$ enters cells through an ionophore for
    a. $PO_4^{3-}$
    b. $Mg^{2+}$
    c. $K^+$
    d. $Cl^-$
    e. $H^+$

18. Insulin is produced by the pancreas in response to increasing the glucose concentration of the blood. What is the most likely action of insulin?
    a. Increases glucose transport proteins
    b. Decreases active transport
    c. Increases bacteriorhodopsin
    d. Decreases phosphoenolpyruvate system
    e. All of the above

# Chapter 5

Use these graphs to answer questions 19-20.

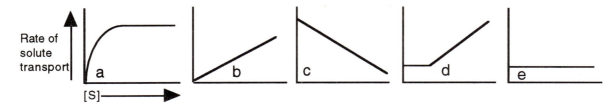

19. Which of the graphs is the best representation of diffusion of the substance into a cell?

20. Which of the graphs is the best representation of facilitated diffusion of the substance into a cell?

21. Plant guard cells have a higher [Cl⁻] than the outside environment. The most likely explanation for this is
    a. Water leaves the cell by osmosis.
    b. Cl⁻ is diffusing through transmembrane channels.
    c. Cl⁻ diffusion is aided by a mobile carrier.
    d. An active transport symport transports Cl⁻/H⁺.
    e. None of the above

## PROBLEMS

1. *Escherichia coli* bacteria adhere to right-side out ghosts but not to inside out ghosts. Moreover, *E. coli* bacteria do not adhere to cells if mannose is in solution around the cells. Explain how these bacteria adhere to cells.

2. You isolate three enzymes from snake venom and subject red blood cells to treatment with each enzyme.

    | Enzyme | Hemolysis |
    | --- | --- |
    | Protease | No |
    | Phosphatidylserine deaminase | No |
    | Phospholipase | Yes |

    In light of what you know about the plasma membrane, explain the results.

3. Both salt (NaCl) and sugar (sucrose) are used as food preservatives. How do they work to preserve foods? Compare the osmotic pressure of 10% NaCl and 20% sucrose.

4. Red blood cells were placed in isotonic solutions of different substances as shown below. How do you account for the results?

# Chapter 5

| Substance | Lysis? | Time to lyse |
|---|---|---|
| $CaCl_2$ | No | – |
| Ethylene glycol | Yes | 1 sec. |
| Fructose | Yes | 2 hr. |
| Glucose | Yes | 2 hr. |
| Glycerol | Yes | 1 min. |
| $NaHCO_3$ | No | – |
| $NaNO_3$ | No | – |
| Urea | Yes | 1 sec. |

5. Plant cells were placed in hypertonic solutions of the following alcohols. Explain the results and suggest relative partition coefficients of each alcohol.

| Alcohol | Plasmolysis? | Time for deplasmolysis |
|---|---|---|
| Ethanol | Yes | 1 min. |
| Ethylene glycol | Yes | 2 min. |
| Glycerol | Yes | – |
| Methanol | No | – |
| Propanol | Yes | 2 min. |
| Propylene glycol | Yes | 15 min. |
| Urea | Yes | 2 min. |

6. Patients with hereditary spherocytosis have fragile red blood cells that can't maintain their biconcave shape. What membrane component are they missing?

7. *Brassica napus* or canola is a source of unsaturated fats. How might the lipid composition change if Canola was grown in temperate compared to subtropical latitudes?

# ANSWERS TO STUDY QUESTIONS

| | | | | | | | |
|---|---|---|---|---|---|---|---|
| 1. a | | 2. b | | 3. b | | 4. c |
| 5. b | | 6. a | | 7. b | | 8. c |
| 9. c | | 10. b | | 11. a | | 12. b |
| 13. b | | 14. c | | 15. d | | 16. b |
| 17. a | | 18. a | | 19. b | | 20. a |
| 21. d | | | | | | |

# CHAPTER 6

## *The Cell Surface and Cellular Communication*

### 👉 LEARNING OBJECTIVES

Be able to
1. Identify the contributions of Connan, Nernst, Goldman, Galvani, Hodgkin and Huxley, and Neher and Sakmann to our understanding of the membrane potential.
2. Describe the action potential of a nerve cell.
3. Describe a voltage-gated channel.
4. Define and give an example of each of the following: signaling molecule, hormone, neurotransmitter.
5. Describe signal transduction in each of the following:
    a. Ion channel receptor
    b. Receptors linked to G proteins
    c. Catalytic receptors
6. Differentiate between stimulating and inhibiting G proteins.
7. Describe the action of the second messengers: cAMP, $IP_3$, DAG, and cGMP.
8. Describe the roles of $Ca^{2+}$, calmodulin, and protein kinases in signal transduction.
9. Explain the mechanism by which G proteins causes different responses in *one* cell and different responses in *different* cells.
10. Explain how receptors are classified.
11. Describe three ways in which signaling responses are stopped.
12. Identify the general chemical composition and function of the extracellular matrix.
13. Compare and contrast the following:
    a. Collagen               b. Elastin
    c. Fibronectin            d. Laminin
    e. Integrin               f. Glycocalyx
    g. N-CAM                  h. Cadherin
    i. Lectin
14. Compare and contrast the following: tight junction, desmosome, adherens junction, and gap junction.
15. Diagram the molecular structure of a plant cell wall, include the middle lamella, primary wall, and secondary wall.
16. Compare the following pairs:
    a. Plant and bacterial cell walls
    b. Animal glycocalyx and bacterial capsule
    c. Gap junction and plasmodesmata
    d. Plasmodesmata and porin

### 👉 CHAPTER OVERVIEW

   **I. THE MEMBRANE POTENTIAL, pp. 195-202**
   1. Unequal distribution of ions inside and outside the cell generates a voltage across the membrane called the membrane potential.
   2. The membrane potential is between -20 and -300 mV. The minus sign indicates the inside of the cell is negatively charged relative to the outside.

A. **The Donnan Equilibrium and Nernst Equation Describe How the Passive Distribution of Ions Influences the Membrane Potential, pp. 196-197**
1. Cells carry a net negative charge because of organic molecules.
2. An ion is at electrochemical equilibrium when the effects of the concentration driving diffusion in one direction equal the effects of the electrical force driving diffusion in the opposite direction.
3. The equilibrium potential is the membrane potential ($\Delta\psi$) at which an ion will not move.
4. The equilibrium or Nernst potential for a cation = $\Delta\psi = 58 \log_{10} \frac{[C^+]_{out}}{[C^+]_{in}}$
5. The equilibrium or Nernst potential for an anion = $\Delta\psi = 58 \log_{10} \frac{[C^-]_{in}}{[C^-]_{out}}$

B. **Active Transport Influences the Membrane Potential, pp. 197-198**
1. Active transport maintains ion concentrations that are not equal to the electrochemical equilibrium.

C. **The Goldman Equation Takes into Account the Contributions of Multiple Ions to the Membrane Potential, p. 198**
1. An electrochemical equilibrium value can be determined for each ion and they can differ from each other and from the actual membrane potential because the membrane is not equally permeable to all ions.
2. The Goldman equation shows that the value of the membrane potential is determined by a membrane's permeability to different ions.

D. **Changes in Membrane Potential Act as a Signaling Device in Nerve and Muscle Cells, pp. 198-199**
1. In the 18th century, Galvani observed that a muscle will contract if its nerve is stimulated electrically.
2. Von Helmholtz calculated that an electrical signal travels through a nerve at ~40 m/sec.
3. Hodgkin and Huxley and Curtis and Cole discovered the action potential: A polarized nerve axon becomes depolarized during stimulation, then the membrane becomes hyperpolarized.

E. **The Action Potential Is Produced by Changes in Membrane Permeability to $Na^+$ and $K^+$, pp. 199-200**
1. Depolarization is caused by increased permeability to $Na^+$; $Na^+$ diffuses into the cell.
2. Hodgkin and Huxley developed the voltage clamp technique in which one electrode monitors the membrane potential and a second electrode delivers electrical current.
3. Repolarization is due to an increase in permeability to $K^+$; $K^+$ diffuses out of the cell.

F. **Voltage-Gated Ion Channels Control the Movement of $Na^+$ and $K^+$ during an Action Potential, pp. 200-202**
1. A single ion channel opening and closing can be measured using the patch clamp technique.
2. Voltage-gated channels are controlled by changes in the membrane potential.

3. Depolarization of a nerve cell causes the Na$^+$ channel to open; the Na$^+$ influx depolarizes the membrane more, opening more Na$^+$ channels.
4. The K$^+$ channels open more slowly allowing an efflux of K$^+$ and repolarization of the membrane.

G. **Inhibitors of Ion Transport Can Block the Na$^+$ and K$^+$ Channels, p. 202**
1. Tetrodotoxin and saxitoxin bind to the outer surface of Na$^+$ channels to block movement of Na$^+$ into the cell.
2. Lipid-soluble anesthetics become inserted into the lipid bilayer and increase membrane fluidity; this interferes with opening Na$^+$ channels.
3. Tetraethylammonium ion blocks K$^+$ channels.

H. **The Permeability of Voltage-Gated Ion Channels Is Controlled by Sequential Conformational Changes, p. 202**
1. Changes in membrane potential cause conformational changes in Na$^+$ channels.
2. In unstimulated cells, the channel is closed.
3. Depolarization causes the channel to open.
4. The channel then assumes an inactivated conformation.

II. **AN OVERVIEW OF SIGNALING MOLECULES AND THEIR RECEPTORS pp. 202-208**
1. Receptors on membranes receive signaling molecules from other cells.
2. A signal molecule causes a conformational change in a receptor.

A. **Hormones Are Transported by the Circulatory System to Target Cells Located throughout the Body, pp. 203-205**
1. Hormones are signaling molecules secreted into the circulatory system by endocrine cells.
2. The three categories of hormones are protein and peptide hormones, steroid hormones, and amino acid derivatives.

B. **Neurotransmitters Are Released into Synapses by Nerve Cells, p. 205**
1. Neurotransmitters include amines: acetylcholine, norepinephrine, dopamine, serotonin, and histamine; amino acids: glutamate, GABA, and glycine; and peptides: enkephalin, β-endorphin, and ATP.

C. **Local Mediators Act on Neighboring Cells, p. 205**
1. Local mediators are secreted by a variety of cells and act on cells within a few millimeters.
2. Regulation by local mediators is called paracrine.
3. Growth factors are proteins that regulate growth and division of cells in multicellular organisms.
4. Lymphokines are proteins that control the development and behavior of immune system cells.
5. Prostaglandins are lipids that act in the cells in which they are made (autocrine) and on neighboring cells; prostaglandins produce a range of effects including inducing contraction or relaxation of smooth muscles.

D. **Hormones, Neurotransmitters, and Local Mediators Transmit Signals by Binding to Receptors, pp. 205-206**
1. Most signal molecules bind to a receptor on the plasma membrane which acts by signal transduction to trigger events inside the cell.

E. **Receptors Are Identified by Testing Their Ability to Bind to Radioactive Signaling Molecules, pp. 206-208**
   1. Binding between a signaling molecule and receptor exhibits:
      a. Specificity between ligand (signaling molecule) and receptor.
      b. High affinity binding or a low dissociation constant ($K_d = 10^{-8}$ to $10^{-11}$).
      c. Saturation kinetics. Plotting the concentration of ligand bound to receptor against the ratio of bound to free ligand is a Scatchard plot.
      d. Reversibility
      e. Physiological response proportional to the number of receptor molecules bound to ligand.

F. **Most Receptors for External Signaling Molecules Are Located in the Plasma Membrane, p. 208**
   1. Most receptors are integral proteins with their binding sites on the exterior leaflet.

III. **ION CHANNEL RECEPTORS, pp. 208-211**

   A. **The Nicotinic Acetylcholine Receptor Has Been Isolated and Purified, pp. 208-210**
      1. Receptors that function as ion channels are called ion channel receptors.
      2. Acetylcholine can bind to different kinds of ion channel receptors including the nicotinic acetylcholine receptor.
      3. An action potential causes the release of acetylcholine from the end of an axon. The acetylcholine diffuses across the synapse to bind to receptors on an adjacent muscle cell.
      4. This triggers depolarization of the target cell.

   B. **The Nicotinic Acetylcholine Receptor Acts as a Neurotransmitter-Gated Ion Channel, pp. 210-211**
      1. The receptor acts as a $Na^+$ channel in a neurotransmitter-gated ion channel.
      2. Binding of acetylcholine causes a conformational change that causes the channel to open.

IV. **RECEPTORS LINKED TO G PROTEINS, pp. 211-221**
   1. G proteins bind GTP.

   A. **Cyclic AMP Functions as a Second Messenger, pp. 211-212**
      1. Glucagon or epinephrine stimulate adenylyl cyclase which produces cAMP from ATP.
      2. cAMP is broken down by phosphodiesterase.
      3. Signal molecules are first messengers; a second messenger transmits the signal to the inside of a cell.

   B. **G Proteins Transmit Signals between Plasma Membrane Receptors and Adenylyl Cyclase, pp. 212-213**
      1. A hormone binds to its receptor; one cell may have receptors for a variety of hormones.
      2. The receptor binds $G_s$ protein which releases GDP and picks up GTP.
      3. The α subunit of $G_s$ dissociates and activates adenylyl cyclase.
      4. When GTP → GDP on the $G_s$ protein, G protein subunits reassociate and adenylyl cyclase is no longer stimulated.

## Chapter 6

**C. Some G Proteins Inhibit Rather Than Stimulate Adenylyl Cyclase, p. 214**
1. A hormone can increase cAMP or decrease cAMP depending on the receptor.
2. Epinephrine stimulates cAMP when it binds to $\beta_1$-adrenergic receptors in heart muscle; and decreases cAMP when it binds to $\alpha_2$-adrenergic receptors on smooth muscle.
3. The $\alpha$ subunit of $G_i$ protein results in inhibition of adenylyl cyclase; the $\beta$ and $\gamma$ subunits combine with the $\alpha$ subunit of $G_s$ to prevent its activation by $G_s$.

**D. G Proteins Are Disrupted by Bacterial Toxins That Cause Cholera and Whooping Cough, pp. 214-215**
1. Cholera toxin catalyzes the transfer of an ADP-ribose from $NAD^+$ to the $\alpha$ subunit of $G_s$; ADP ribosylation inhibits the hydrolysis of GTP by $G_s$ leaving adenylyl cyclase activated. Diarrhea results from increased secretion of $Na^+$ and water into the intestinal tract.
2. Whooping cough toxin catalyzes ADP ribosylation of $G_i$ leading to increased production of cAMP and fluid secretion into the lungs.

**E. Cyclic AMP Activates Protein Kinase A, p. 215**
1. Protein kinases catalyze the transfer of a phosphate group from ATP to an amino acid in a protein molecule (protein phosphorylation) to regulate the activity of the protein. (Protein kinase A is a protein-serine/threonine kinase.)
2. Binding cAMP causes dissociation of protein kinase subunits. This results in phosphorylation of a protein.

**F. Different Proteins Are Phosphorylated in Different Target Cells by Protein Kinase A, p. 216**
1. In liver and muscle cells, phosphorylase kinase is activated. It activates glycogen phosphorylase, glycogen synthase, protein phosphatase-1, and protein phosphatase inhibitor-1.
2. In fat cells, protein kinase A activates hormone-sensitive lipase.

**G. Advantages of the Cyclic AMP System Include Amplification and Flexibility, pp. 217-218**
1. Many molecules of $G_s$ are activated by one signal and adenylyl cyclase produces hundreds of cAMP per second.
2. Each step in cAMP-mediated signal transduction can be regulated by alternative mechanisms.

**H. G Proteins Can Alter the Permeability of Ion Channels, p. 218**
1. Muscarinic acetylcholine receptors of the heart bind acetylcholine which activates $G_i$. The $\alpha$ subunit opens $K^+$ channels to hyperpolarize the plasma membrane.

**I. G Proteins Are Involved in the Phosphoinositide Signaling Pathway, p. 218**
1. Signal molecules bind membrane receptors activating $G_p$ which activates phospholipase C which cleaves $PIP_2$ into $IP_3$ and DAG.

**J. Inositol Trisphosphate ($IP_3$) Is a Second Messenger That Mobilizes Calcium Ions, pp. 219-221**
1. $IP_3$ serves as a second messenger by binding to the ER to release stored $Ca^{2+}$.

# Chapter 6

      2. $Ca^{2+}$ binds calmodulin; calmodulin undergoes a conformational change to activate other target proteins.
      3. Calmodulin-dependent multiprotein kinase catalyzes the phosphorylation of many different proteins.
      4. $Ca^{2+}$-calmodulin stimulates adenylyl cyclase and phosphodiesterase.

   **K. Diacylglycerol (DAG) Is a Second Messenger That Activates Protein Kinase C, p. 221**
      1. DAG stimulates protein kinase C.
      2. Phorbol esters can stimulate protein kinase C.
      3. $Na^+$-$H^+$ antiport is activated by protein kinase C.

**V. CATALYTIC RECEPTORS, pp. 221-224**

   **A. Many Growth Factors Interact with Receptors That Function as Protein-Tyrosine Kinases, pp. 221-222**
      1. PDGF and EGF bind to a receptor protein kinase called protein-tyrosine kinase.
      2. Tyrosine kinases exhibit autophosphorylation.

   **B. Receptors Exhibiting Guanylyl Cylase Activity Produce the Second Messenger Cyclic GMP, p. 222**
      1. Atrial natriuretic peptide activates guanylyl cyclase to produce cGMP.
      2. cGMP enhances excretion of $Na^+$ and water and relaxes smooth muscle.

   **C. Receptors Are Grouped into Superfamilies, pp. 222-223**
      1. Single-pass receptors consist of a single transmembrane α helix; e.g., protein-tyrosine kinases and guanylyl cyclase.
      2. Seven-pass receptors contain seven transmembane α helices; e.g., muscarinic receptor.
      3. Multisubunit receptors resemble the nicotinic acetylcholine receptor.

   **D. Signaling Responses Are Terminated by Reducing the Concentration of Signaling Molecules or Active Receptors, pp. 223-224**
      1. Signal molecules can be **removed** by enzymatic hydrolysis; $Ca^{2+}$ is pumped from the cell.
      2. Receptors can be **desensitized**; in homologous desensitization, receptors bound to the signal molecule are affected; in heterologous desensitization, receptors are desensitized by any signal that increases cAMP.
      3. **Receptor down-regulation** reduces the number of receptors.

**VI. THE EXTRACELLULAR MATRIX, pp. 224-233**
      1. Cells secrete macromolecules called the extracellular matrix.
      2. In humans, fibroblasts are primarily responsible for the extracellular matrix.
      3. The matrix influences cell shape and motility, growth and division, and specialized cell characteristics.
      4. Matrix molecules include: glycosaminoglycans and proteoglycans, structural proteins, and adhesive proteins.

Chapter 6

A. **The Ground Substance of the Extracellular Matrix Is Formed by Glycosaminoglycans and Proteoglycans, pp. 224-225**
   1. The ground substance of the matrix consists of glycosaminoglycans that are composed of repeating disaccharide units containing one amino sugar and one $SO_4^-$ or $COO^-$; e.g., chondroitin sulfate and hyaluronate.
   2. Most glycosaminoglycans are bound to proteins to form proteoglycans.
   3. Glycosaminoglycans form a hydrated gel that affects the shape, strength, and rigidity of a tissue.

B. **Collagen Is Primarily Responsible for the Strength of the Extracellular Matrix, pp. 226-228**
   1. Collagen is a rigid triple helix; 25% glycine and 25% hydroxyproline and hydroxylysine.
   2. Ascorbic acid maintains the activity of prolyl hydroxylase which produces hydroxyproline; ascorbic acid deficiency results in unstable collagen and is called scurvy.
   3. Collagen exists as banded fibrils and unbanded filamentous networks. Banded fibrils aggregate to form collagen fibers.

C. **Collagen Is Produced from a Precursor Molecule Called Procollagen, pp. 228-229**
   1. Procollagen is formed in the ER.
   2. Procollagen is secreted into the extracellular space and converted to collagen.

D. **Elastin Imparts Elasticity and Flexibility to the Extracellular Matrix, p. 229**
   1. Elastic fibers made of elastin give flexibility to tissues such as arteries.

E. **Fibronectin Binds Cells to the Matrix and Guides Cellular Migration, p. 230**
   1. Arg-Gly-Asp binds fibronectin to the cell surface; other domains bind fibronectin to collagen, heparin, and fibrin.
   2. Fibronectin attaches platelets to fibrin and guides the migration of immune system cells.

F. **Laminin Binds Cells to the Basal Lamina, p. 231**
   1. The basal lamina is 50 nm thick and separates epithelial cells from underlying supporting tissue; it provides structural support and maintains tissue organization.
   2. The basal lamina prevents passage of connective tissue cells into the epithelium but allows passage of immune system cells.

G. **Integrin Receptors Bind to Fibronectin, Laminin, and Other Matrix Constituents, pp. 231-232**
   1. Integrins are cell surface molecules that bind to the extracellular matrix.

H. **The Glycocalyx Is a Carbohydrate-Rich Zone Located at the Periphery of Many Animal Cells, pp. 232-233**
   1. A carbohydrate-rich zone called the glycocalyx surrounds animal cells.
   2. The attached glycocalyx is an integral part of the plasma membrane; the unattached glycocalyx consists of glycoproteins and proteoglycans that are also part of the extracellular matrix.

## Chapter 6

**VII. CELL-CELL RECOGNITION AND ADHESION, pp. 233-236**
   1. In 1907, Wilson observed that cells recognize and selectively bind to each other.

   **A. N-CAMs and Cadherins Are Plasma Membrane Glycoproteins that Mediate Cell-Cell Adhesion, pp. 233-234**
      1. N-CAM on one cell binds to N-CAM on another cell.
      2. $Ca^{2+}$ produces a conformational change in cadherin that allows cadherin to mediate cell adhesion.

   **B. Carbohydrate Groups Participate in Cell-Cell Recognition and Adhesion, pp. 234-236**
      1. The amount of sialic acid in N-CAM regulates cell-cell adhesion.
      2. Galactose is thought to be involved in cell-cell adhesion.
      3. Proteins called lectins bind to specific sugars on cells.
      4. Glycophorins are involved in cell recognition.
      5. In the ABO blood group system, type A cells have glycophorin chains ending in N-acetylgalactosamine and chains from type B cells end in galactose.
      6. Sialic acid in RBC glycophorin decreases as a cell ages; the absence of sialic acid is used to recognize old RBCs which are destroyed.

**VIII. CELL JUNCTIONS, pp. 236-243**

   **A. Tight Junctions Create a Permeability Barrier Across a Layer of Cells, pp. 236-238**
      1. Tight junctions prevent movement of molecules and prevent lateral diffusion of proteins through the cell membrane.
      2. The membranes of adjacent cells are fused at a tight junction.

   **B. Plaque-Bearing Junctions Stabilize Cells Against Mechanical Stress, pp. 238-240**
      1. Plaque-bearing junctions provide mechanical strength.
      2. Plaque-bearing junctions connect the cytoskeleton of adjacent cells.
      **3. Plaque-bearing Junctions Associated with Intermediate Filaments Are Called Desmosomes, pp. 238-239**
         a. Desmosomes are 10-nm bundles of intermediate filaments common in skin and intestines.
         b. Cells are joined by membrane glycoproteins.
         c. Cells use intermediate filaments to attach to the basal lamina at hemidesmosomes.
      **4. Plaque-bearing Junctions Associated with Actin Filaments Are Called Adherens Junctions, pp. 239-240**
         a. Epithelial cells can have adhesion belts covering an entire cell.
         b. Adherens junctions use 6-nm bundles of actin filaments and transmembrane proteins to join cells.
         c. Cells attach to the extracellular matrix by focus adhesion.

   **C. Gap Junctions Permit Small Molecules to Move from One Cell to Another, pp. 240-243**
      1. Plasma membranes of two adjacent cells are aligned and separated by a 3 nm gap.
      2. The membranes at the gap are covered with cylindrical connexons.

## Chapter 6

    2. The membranes at the gap are covered with cylindrical connexons.
    3. Connexons in adjacent cells are connected to form a channel through which water-soluble molecules can pass.
    4. Gap junctions open and close with the intracellular $Ca^{2+}$, cAMP, and pH and membrane potential.

### IX. PLANT CELL SURFACE, pp. 243-248

**A. Plant Cell Walls Provide a Supporting Framework for Intact Plants, p. 243**
    1. Enzymes associated with the wall can degrade extracellular nutrients.

**B. The Plant Cell Wall Is Constructed from Cellulose, Hemicellulose, Pectin, Lignin, and Glycoproteins, pp. 243-245**
    1. Cellulose microfibrils are twisted into macrofibrils to make up the cell wall.
    2. Hemicellulose (consisting of xylose, arabinose, mannose, or galactose) coats cellulose microfibrils.
    3. Pectin (a polymer of galacturonic acid) is a hydrated gel that binds the walls of adjacent cells.
    4. Lignins are polymerized aromatic alcohols located between cellulose fibrils. Lignin accounts for up to 25% of the dry weight of woody plants.
    5. Glycoproteins such as extensins form a crosslinked network with cellulose fibrils.

**C. The Plant Cell Wall Is Synthesized in Several Discrete Stages, pp. 245-246**
    1. The middle lamella composed of pectin is produced first.
    2. The cellulose primary cell wall is laid down next. Cellulose microfibrils are made by cellulose-synthesizing enzymes located in rosettes in the plasma membrane.
    3. Woody plants add a secondary cell wall inside the primary wall after cell growth has stopped. The cellulose and lignin of the secondary wall are densely packed.

**D. Plasmodesmata Permit Direct Cell-Cell Communication through the Plant Cell Wall, pp. 246-247**
    1. Plasmodesmata are channels in the cell wall that are lined by plasma membrane and contain ER tubules, so the cytoplasm, plasma membrane, and ER of adjacent cells are in contact.
    2. Plasmodesmata are located in pit fields.
    3. Plasmodesmata may be regulated by intracellular $Ca^{2+}$.

**E. Chemical and Electrical Signaling Occur in Plants as Well as Animals, pp. 247-248**
    1. Plant hormones are small and freely pass through the cell wall.
    2. Changes in the action potential (due to permeability changes to $Ca^{2+}$ and $Cl^-$) of plant cells is used in cell-cell signaling.

### X. THE BACTERIAL CELL SURFACE, pp. 248-252
    1. The bacterial cell envelope consists of the plasma membrane and the layers external to it.
    2. The wall protects the cell from osmotic lysis.
    3. The wall is responsible for the shape of a bacterium: i.e., cocci, bacilli, spirilla, and spirochetes.

# Chapter 6

    4. Bacteria are classified by the reactions of the cell walls to the Gram stain procedure: Gram-positive cells are not decolorized by alcohol, Gram-negative cells are decolorized by alcohol.

A. **Gram-Positive Bacteria Have a Thick Cell Wall Made of Murein and Teichoic Acids, pp. 248-250**
1. Peptidoglycan consists of repeating disaccharide units of N-acetylglucosamine and N-acetylmuramic acid (NAM). The NAM of one chain is connected to a parallel chain by a peptide bridge.
2. Peptidoglycan is hydrolyzed by lysozyme; wall-less bacteria are called protoplasts.
3. Teichoic acids are linked to the peptidoglycan skeleton.

B. **Gram-Negative Bacteria Have a Thin Murein Wall plus an Outer Membrane, pp. 250-251**
1. The inner leaflet of the outer membrane contains phospholipids.
2. The outer leaflet consists of lipopolysaccharides.
3. The outer membrane repels hydrophobic molecules.
4. The lipopolysaccharide is an endotoxin that causes an inflammatory reaction (e.g., fever, capillary dilation) in humans.
5. Porin proteins form pores in the outer membrane for the diffusion of small water-soluble molecules.

C. **Synthesis of the Bacterial Cell Wall Is Blocked by Penicillin, p. 251**
1. The cell wall is synthesized by the plasma membrane.

D. **The Periplasmic Space Is a Unique Compartment outside the Plasma Membrane of Gram-Negative Bacteria, p. 251**
1. The area between the plasma membrane and the outer membrane is the periplasmic space.
2. The periplasm contains detoxifying enzymes, binding proteins, and hydrolytic enzymes.
3. Similar enzymes are loosely bound to the surface of or are secreted by Gram-positive cells.

E. **Capsules Can Be Produced by Gram-Positive and Gram-Negative Bacteria, pp. 251-252**
1. A capsule surrounds the cell wall and is attached to the outermost layer of the wall.
2. Capsules can be made of polysaccharides or polypeptides (made of D-amino acids).
3. Capsules can protect bacteria from destruction by cells of the host's immune system.

# KEY TERMS

| | | |
|---|---|---|
| action potential | adenylyl cyclase | adherens junction |
| adhesion belt | ADP ribosylation | adrenergic receptor |
| AMP-dependent protein kinase | atrial natriuretic peptide | autocrine |
| axon | bacilli | basal lamina |
| cadherin | calmodulin-dependent multiprotein kinase | cAMP |

- capsule
- cell junction
- cocci
- connexin
- desmosome
- elastin
- equilibrium potential
- fibroblast
- G protein
- glycocalyx
- glycoprotein
- Gram-negative
- hemicellulose
- hyperpolarized
- intermediate filament
- lectin
- lipopolysaccharide
- membrane potential
- muscarinic Ach receptor
- neurotransmitter
- outer membrane
- pectin
- periplasmic space
- phosphoinositide signaling
- plasmodesmata
- procollagen
- protein phosphatase inhibitor-1
- protein-serine/threonine kinase
- protoplast
- secondary cell wall
- spirochete
- tight junction
- catalytic receptor
- cellulose
- collagen
- connexon
- diacylglycerol
- electrochemical equilibrium
- extensin
- fibronectin
- G protein
- glycogen phosphorylase
- glycosaminoglycan
- Gram-positive
- hemidesmosome
- inositol trisphosphate
- ion channel receptor
- ligand
- local mediator
- middle lamella
- N-CAM
- neurotransmitter-gated ion channel
- paracrine
- penicillin
- phorbol esters
- phospholipase C
- porin
- protein kinase A
- protein phosphatase-1
- protein-tyrosine kinase
- receptor
- signal transduction
- synapse
- voltage-gated channel
- cell envelope
- cGMP
- collagen fibril
- depolarization
- dissociation constant
- endocrine
- extracellular matrix
- focal adhesion
- gap junction
- glycogen synthase
- Goldman equation
- growth factor
- hormone
- integrin
- laminin
- lignin
- lymphokine
- murein
- Nernst potential
- nicotinic Ach receptor
- patch clamp technique
- peptidoglycan
- phosphodiesterase
- plaque-bearing junction
- primary cell wall
- protein kinase C
- protein phosphorylation
- proteoglycan
- rosette
- spirilla
- teichoic acid

# Chapter 6

# 👉 KEY FIGURE

Label the following: hormone, receptor, $G_s$ protein, adenylyl cyclase, GTP, GDP, cAMP and show how signal transduction works.

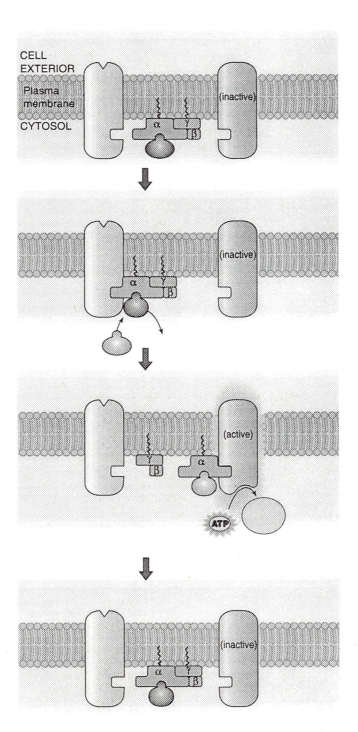

# Chapter 6

# 👉 STUDY QUESTIONS

1. Which of the following pairs is **not** correctly matched?
    a. Porin — ionophore
    b. Plasmodesmata — gap junction
    c. Plant cell wall — bacterial cell wall
    d. Tight junction — middle lamella
    e. None of the above

2. Treatment of an infection by a Gram-negative bacterium results in a life-threatening condition called septic shock. This is due to the
    a. Release of cell wall lipopolysaccharides from the dead cells.
    b. Production of an exotoxin.
    c. Inhibition of G protein.
    d. Release of peptidoglycan.
    e. None of the above

3. Cocaine inhibits the reabsorption of the neurotransmitter norepinephrine. What effect does this have on a nerve cell?
    a. Prevents depolarization.
    b. Causes polarization.
    c. Causes hyperpolarization.
    d. Causes action potentials.
    e. None of the above

4. A cancer-causing gene makes G [$G_s$] protein permanently bind to GTP, this results in
    a. Continual production of cAMP.
    b. Continual production of GDP.
    c. Inhibition of cGMP production.
    d. Inhibition of adenylyl cyclase.
    e. None of the above

5. The steps for an action potential are listed below. Which is the first step?
    a. Depolarization
    b. $K^+$ channels open
    c. $Na^+$ channels open
    d. Polarization
    e. Repolarization

6. The steps in G protein-mediated signal transduction are listed below. What is the third step? [s]
    a. G [$_s$] protein ~~subunits reassociate~~ [binds to receptor].
    b. Hormone binds to its receptor.
    c. $G_s$ protein ~~picks up~~ [binds to] GTP.
    d. $G_s$ activates adenylyl cyclase.
    e. GTP → GDP.

81

# Chapter 6

7. Canine kidney cell cultures develop invasive cancer cell characteristics when they are treated with antibodies against E-cadherin. This indicates that the role of cadherin is
   a. Signal transduction.
   b. Cell communcation.
   c. Initiating an action potential.
   d. Cell migration.
   e. None of the above

8. *E. coli* 0157:H7 bacteria produce an enterotoxin that allows $G_s$ to bind GTP but inhibits the hydrolysis of GTP. The effect is
   a. cAMP cannot be formed.
   b. Adenylyl cyclase will continuously produce cAMP.
   c. The cell will not respond to a signal.
   d. Second messengers will not be formed.
   e. None of the above

9. The hormone glucagon stimulates the breakdown of stored glycogen in liver and muscle cells by the following enzymes. Which is the first enzyme that must be activated?
   a. Protein kinase A
   b. Phosphorylase kinase
   c. Glycogen phosphorylase
   d. Protein phosphatase
   e. Protein phosphatase inhibitor-1

10. Which of the following is **not** a characteristic of the bonding between a signaling molecule and its receptor?
    a. The reaction is specific.
    b. The dissociation constant for the reaction is low.
    c. The reaction is irreversible.
    d. The bonding brings about a physiological response.
    e. None of the above

11. Cancerous liver cells can be restored to their normal growth by inserting the gene for E-cadherin. This indicates that cadherin is involved in
    a. Formation of the extracellular matrix.
    b. Opening voltage-gated channels.
    c. Activation of cAMP.
    d. Cell migration.
    e. None of the above

12. Fruit juices are made by adding pectinase to crushed fruit. The function of the pectinase is to
    a. Remove the plant cell walls.
    b. Lyse the cells.
    c. Separate the plant cells.
    d. Remove the cell membrane.
    e. None of the above

# Chapter 6

For questions 13-18: Compare the paired choices for each question and answer
- a if A is greater than B
- b if B is greater than A
- c if the two are nearly equal

> For example: The ease of seeing bacteria
> A. With a microscope
> B. Without a microscope        Answer:    a

13. The requirement for $Ca^{2+}$ to activate
    A. Calmodulin
    B. G protein

14. The amount of cAMP that is made when
    A. The α subunit of $G_s$ is bound to adenylyl cyclase
    B. The α subunit of $G_s$ is bound to the β and γ subunits

15. In cancer cells, the number of adherens junctions that are
    A. Normal
    B. Abnormal

16. The necessity for $Ca^{2+}$ for cell adhesion mediated by
    A. N-CAM
    B. Cadherin

17. The membrane potential when a cell is
    A. Depolarized
    B. Repolarized

18. The equilibrium potential required to balance the tendency
    A. of $K^+$ to leave a cell if the concentration of $K^+$ inside the cell is 100 mM and outside the cell is 10 mM
    B. of $Cl^-$ to enter a cell if the concentration of $Cl^-$ inside the cell is 52 mM and outside the cell is 520 mM

19. $Ca^{2+}$ seems to be necessary for viral infections. The role of $Ca^{2+}$ is most likely
    a. To activate calmodulin.
    b. To activate N-CAM.
    c. To mediate cell-to-virus adhesion.
    d. To form tight junctions.
    e. None of the above

20. Developing frog embryos treated with anti-fibronectin antibodies
    a. Develop normallly.
    b. Stop cell division.
    c. Have abnormal cell migration.
    d. Don't exhibit action potentials.
    e. None of the above

21. During the day, plant guard cells become turgid which opens a pore (stomate) between adjacent guard cells. The following steps are required for stomatal opening. Which is the fifth step?
    a. $H^+$ pumped out by ATPase—hyperpolarization
    b. $K^+$ influx—repolarization
    c. Light—signal
    d. Water enters—turgidity
    e. $K^+$ channels open—voltage-gated channel

# PROBLEMS

1. It has been said that plants are not multicellular but are actually one multinucleated cytoplasmic continuum. What structural characteristics of plants would lead to this conclusion? Make an argument that plants are (or are not) multicellular organisms.

2. *Streptococcus pyogenes* bacteria produce the enzyme hyaluronidase. This enzyme contributes to the bacterium's virulence (ability to cause disease) by enabling the bacteria to spread through the human body. How does this enzyme allow the bacterium to spread?

3. You prepare a cell sample for light microscopy and stain it with acid methylene blue ($H^+$ methylene blue) and observe that the background stains and not the cell. You adjust the pH of the methylene to produce an alkaline compound (methylene blue $Cl^-$) and observe that the cells stained this time. Explain what happened.

4. In 1986, Rita Levi-Montalcini received a Nobel Prize for her discovery. She observed that when mouse sarcoma (cancer) cells are grown with chick embryo cells, the chick cells develop into nerve cells. What did Montalcini discover? (Growth factors; specifically nerve growth factor)

5. Marine snail larvae are free-floating in the ocean, but must settle down on the appropriate alga in order to mature into adults and reproduce. Marine algae produce a GABA-like protein. Propose a mechanism by which the larvae can settle on an alga.

6. In 1899, Jacques Loeb investigated how development in eggs is activated. He placed unfertilized sea urchin eggs in sea water to which he had added sodium hydroxide to increase the alkalinity. The unfertilized eggs divided once or twice. Later, he succeeded in getting unfertilized eggs to develop into larvae by placing them in sea water with 50 mM $K^+$. Normal seawater is 10 mM $K^+$ and sea urchin eggs contain 300 mM $K^+$. What can you conclude about activation of an egg?

7. The most common hereditary hemophilia is von Willebrand disease (vWD). Patients with vWD form abnormal platelet plugs which result in prolonged bleeding time. Platelets from a patient with vWD do not adhere to vWD endothelial cells. vWD platelets will adhere to normal endothelial cells but not to normal embryonic cells. Explain the defect in vWD.

# ☞ ANSWERS TO STUDY QUESTIONS

| 1. d | 2. a | 3. d | 4. a |
|---|---|---|---|
| 5. a | 6. d | 7. d | 8. b |
| 9. a | 10. c | 11. d | 12. c |
| 13. a | 14. a | 15. a | 16. b |
| 17. a | 18. c | 19. c | 20. c |
| 21. d | | | |

# CHAPTER 7

## *Cytoplasmic Membranes and Intracellular Traffic*

### LEARNING OBJECTIVES

Be able to
1. Define the following terms and provide a function for each organelle:
   a. Endoplasmic reticulum
   b. Microsomes
   c. Golgi complex
   d. Lysosome
   e. Coated pits
2. Differentiate between the following terms:
   a. Smooth ER and rough ER
   b. *cis* Golgi network and *trans* Golgi network
   c. Lysosome and spherosome
   d. Exocytosis and endocytosis
   e. Macroautophagy and microautophagy
   f. Phagocytosis and pinocytosis
   g. Primary lysosome and secondary lysosome
   h. Autolysis and extracellular digestion
   i. Coated pits and receptor-mediated endocytosis
3. Describe the role of the smooth ER in glucose release, fat storage, membrane synthesis, and detoxification.
4. Describe how proteins are targeted for synthesis on the ER.
5. Differentiate between the following terms:
   a. Targeting signal and signal sequence
   b. Signal sequence and signal recognition particle
   c. Stop-transfer sequence and signal sequence
   d. Storage vesicle and transitional vesicle
   e. Constitutive secretion and regulated secretion
6. List three ways proteins are modified in the ER.
7. Identify the functions of the Golgi complex.
8. Describe how lysosomal enzymes are synthesized and packaged into lysosomes.
9. List the major characteristics of lysosomal storage diseases.
10. Outline the steps of pinocytosis. Of phagocytosis.
11. Describe the role of phospholipid translocators and phospholipid transfer proteins in membrane biogenesis.

### CHAPTER OVERVIEW

**I. INTRODUCTION, p. 256**
  1. Claude developed procedures for isolating organelles by subcellular fractionation.
  2. Palade used fractionation and electron microscopy to study the ER and Golgi complex.
  3. De Duve predicted the existence of lysosomes using fractionation.

II. **ENDOPLASMIC RETICULUM, pp. 256-272**
1. Early microscopists observed densely staining regions in cells they called ergastoplasm. In 1945, Porter, Claude, and Fullam observed that the ER consisted of membranous channels, vesicles, and sacs.

A. **The Endoplasmic Reticulum Consists of Two Components, the Smooth ER and the Rough ER, pp. 257-259**
1. The rough ER is composed of large flattened membrane sacs and has ribosomes attached to it. It is found in cells that synthesize proteins to be secreted.
2. The smooth ER is composed of interconnected membrane tubules and lacks ribosomes. It is most pronounced in cells that metabolize lipids, drugs, and toxins.
3. The ER divides the cytoplasm into the cytosol (containing enzymes, ribosomes, and tRNAs) and the cisternal space or ER lumen (between the ER membranes).
5. The divisions created by the ER can have different purposes.
6. Microsomes are vesicles that form from ER fragments created during cell homogenization.

B. **The Smooth ER Is Involved in Releasing Free Glucose from Glycogen, pp. 259-260**
1. Glucose 6-phosphate is hydrolyzed to glucose and $P_i$ in the smooth ER.

C. **The Smooth ER Is Involved in Synthesizing Triacylglycerols and Steroids, p. 260**
1. Triacylglycerols are stored in the ER lumen in adipocytes.
2. Mevalonate is synthesized in the smooth ER; mevalonate is converted into cholesterol in the cytosol; cholesterol is converted into steroid hormones in the ER.

D. **The Smooth ER Synthesizes the Phospholipids Needed for Cellular Membranes, pp. 260-261**
1. Membrane phospholipids are synthesized on the cytosol-facing side of the ER.
2. As new phospholipids are made, the ER grows larger; phospholipid translocators move phospholipids from one half of the membrane to the other.

E. **The Smooth ER Oxidizes Foreign Substances Using Cytochrome P-450, pp. 261-263**
1. The oxidization of substances using $O_2$ is carried out by mixed-function oxidases.
2. An oxidizable substrate binds to cytochrome P-450.
3. The P-450 iron atom is reduced by NADPH.
4. $O_2$ binds to P-450.
5. One O atom oxidizes the substrate, the other forms water while oxidizing the iron atom.
6. The presence of certain substrates (e.g., phenobarbital) causes proliferation of the smooth ER and some of the ER-associated enzymes.
7. Cytochrome P-448 (aryl hydrocarbon hydroxylase) is involved in oxidizing polycyclic hydrocarbons.
8. Cytochrome $b_5$ uses NADH as a coenzyme; it is involved in the desaturation of fatty acids.

F. **The Rough ER Is Involved in Protein Targeting, pp. 263-264**
1. Secretory proteins and lumenal proteins are synthesized on membrane-attached ribosomes and released into the ER lumen for distribution to the Golgi complex, lysosomes, and other vesicles.
2. Integral membrane proteins are synthesized on membrane-attached ribosomes and retained in the lipid bilayer of the ER.
3. Targeting signals determine whether a forming polypeptide will be made on membrane-attached or free ribosomes; they also determine the final destination of the protein.

G. **Signal Sequences Target Proteins to the Endoplasmic Reticulum, pp. 264-265**
1. Proteins are transported into the ER lumen as they are made.
2. Protein synthesis begins on free ribosomes and signal sequences near the N-terminal end of a polypeptide and cause the ribosome to attach to the ER.
3. Preproteins contain the signal sequence of amino acids. The signal sequence is removed by the ER.

H. **Proteins Synthesized on Membrane-Bound Ribosomes Pass Through Protein-Translocating Channels in the ER Membrane, pp. 265-268**
1. Free ribosomes bind to mRNA.
2. After the signal sequence is formed, the SRP binds to the signal sequence.
3. The SRP binds to the ER at the SRP receptor.
4. SRP is released and the ribosome and its polypeptide chain attach to the protein-translocating channel.
5. The growing polypeptide chain emerges into the ER lumen and the signal sequence is removed by signal peptidase.
6. When the polypeptide exits from the channel, the ribosome dissociates and the channel closes.

I. **Integral Membrane Proteins Contain Stop-Transfer Sequences as well as Signal Sequences, p. 268**
1. Membrane proteins are anchored to the membrane bilayers by one or more transmembrane α-helices.
2. For proteins with the N-terminal end facing the lumen: The hydrophobic α-helix acts as a stop-transfer sequence that stops movement of the growing polypeptide through the channel.
3. For proteins with the C-terminal end facing the lumen: The signal sequence acts as the stop sequence.

J. **Insertion or Translocation into the ER Can Also Occur after a Protein Has Been Synthesized, pp. 268-269**
1. Chaperone proteins bind to the reactive surface of polypeptides to protect them from other reactive surfaces before they are inserted into the ER membrane.

K. **Newly Made Proteins Are Modified in the ER by N-Linked Glycosylation, Hydroxylation, and Linkage to Membrane-Bound Glycolipid, pp. 269-270**
1. Most proteins made in the rough ER are glycoproteins.
2. **N-linked glycosylation, p. 269-270**
    a. A core oligosaccharide is attached to the free $NH_2$ of asparagine in the sequence Asn-S-Ser/Thr.

b. The core oligosaccharide is assembled from N-acetylglycosamine, mannose, and glucose on dolichol phosphate embedded in the ER membrane.
3. **Hydrolyxation, p. 270**
   a. Some proteins, e.g., collagen, are hydroxylated at proline and lysine.
4. **Attachment to membrane-bound glycolipid, p. 270**
   a. Lipid-anchored proteins are attached to membrane-bound glycolipid.

L. **Newly Formed Proteins Acquire Their Proper Conformation in the ER, pp. 270-272**
   1. Protein disulfide isomerase catalyzes the formation and breakage of disulfide bonds.
   2. Chaperone proteins, called binding proteins (BiP), stabilize unfolded polypeptides.
   3. A polypeptide may be cleaved to form the protein subunits.

III. **THE GOLGI COMPLEX, pp. 272-283**

   A. **The Golgi Complex Consists of a Stack of Flattened Membrane Cisternae and Associated Vesicles, pp. 272-274**
      1. In 1889, Golgi discovered a threadlike network (Golgi complex) near the nucleus.
      2. The Golgi complex consists of flattened membrane vesicles called cisternae, smaller vesicles, and membrane channels.
      3. The side of the cisternae facing the ER is the *cis* face, the opposite side is the *trans* face.
      4. Thiamine pyrophosphatase and nucleoside diphosphatase occur primarily in the Golgi complex.

   B. **Proteins Synthesized in the Rough ER Are Routed Through the Golgi Complex, pp. 274-277**
      1. Secretory proteins are synthesized in the rough ER, transported through the Golgi complex into secretory vesicles, and secreted.
      2. Palade *et al.* used $^3$H-leucine to study the movement of newly made proteins in guinea pig pancreas cells.
      3. Yeast *sec* mutants are unable to secrete proteins at high temperatures. Many *sec* genes code for G proteins.

   C. **Vesicles Transport Materials Between the Endoplasmic Reticulum and the Golgi Complex, pp. 277-279**
      1. Transitional vesicles bud off from the rough ER and fuse with the *cis* Golgi network.
      2. Proteins with the C-terminal signal Lys-Arg-Glu-Leu (KDEL) are returned to the ER.

   D. **Proteins Are Glycosylated as They Pass through the Golgi Complex, p. 279**
      1. Neutra and Leblond used $^3$H-glucose to demonstrate that glycosylation of secretory proteins occurs in the Golgi complex.
      2. The N-linked oligosaccharides attached to proteins in the ER are modified in the Golgi complex.
      3. Glycosyl transferases attach sugars to the hydroxyl groups of serine, threonine, and hydroxylysine (O-linked glycosylation).

E.  **The Golgi Complex Is Also Involved in Synthesizing Polysaccharides, p. 279**
    1.  The Golgi complex synthesizes hyaluronate in animal cells and hemicelluloses and pectins in plant cells.

F.  **Proteins Are Sorted by the Golgi Complex Using Chemical Signals and Bulk Flow, p. 280**
    1.  Targeting signals on proteins are used to sort proteins in the Golgi complex, e.g., mannose 6-phosphate codes proteins for lysosomes.
    2.  In bulk flow, unlabeled proteins in the ER are transported to the Golgi complex and then into secretory vesicles. The vesicles fuse with the plasma membrane and expel their contents.

G.  **The Golgi Complex Produces Various Kinds of Intracellular Granules, pp. 281-283**
    1.  A variety of storage vesicles are produced by the Golgi complex, e.g., azurophil granules and specific granules of neutrophils.

IV. **CELL SECRETION, pp. 283-286**

   A.  **Secretion Can be Either Constitutive or Regulated, p. 283**
       1.  Constitutive secretion is continuous.
       2.  Regulated secretion occurs in specialized cells, producing specific products.

   B.  **Vesicles Destined for the Constitutive and Regulated Secretory Pathways Can be Distinguished from One Another, pp. 283-284**
       1.  Constitutive products are concentrated in condensing vacuoles or in *trans* flattened cisternae.
       2.  As they bud from the *trans* Golgi network, the vesicles are coated with clathrin.
       3.  Small vesicles fuse with each other to produce secretory vesicles.
       4.  Transport vesicles are used for regulated secretion.
       5.  The low pH of the *trans* Golgi network may function to sort regulated from constitutive products.

   C.  **Cleavage of Precursor Proteins Often Occurs in Immature Secretory Vesicles, p. 284**
       1.  Proinsulin and (probably) a protease are packaged in clathrin-coated vesicles; cleavage of the polypeptide chain in the secretory vesicles produces insulin.
       2.  (Inactive) trypsinogen and chymotrypsinogen are secreted from pancreas cells.

   D.  **Materials Are Expelled from Cells by Exocytosis, pp. 284-285**
       1.  In exocytosis, the membrane of a secretory vesicle fuses to the plasma membrane to release secreted proteins.
       2.  Exocytosis occurs spontaneously for constitutive proteins.
       3.  Exocytosis is triggered by a neurotransmitter or hormone for regulated proteins.
       4.  $Ca^{2+}$, GTP, and ATP are required for exocytosis.
       5.  Exocytosis frequently occurs at a specific cite on the plasma membrane, controlled by microtubules and motor proteins (dynein and kinesis).

E.  **Membrane Components Are Recycled from the Plasma Membrane Back to the Golgi Complex, pp. 285-286**
    1.  Vesicles bud off from the plasma membrane and migrate back to the Golgi complex.
    2.  Vesicles are labeled with t-SNAREs and fuse to target membranes containing the appropriate v-SNAREs.

F.  **Bacterial Cells Secrete Materials Directly through the Plasma Membrane, p. 286**
    1.  In bacteria, secretory proteins contain a signal sequence for binding to the plasma membrane.

V.  **LYSOSOMES AND ENDOCYTOSIS, pp. 286-300**
    1.  De Duve *et al.* identified lysosomes using subcellular fractionation.

   A.  **Acid Phosphatase Is Localized within a Membrane-Enclosed Organelle, pp. 286-287**
       1.  Reagents used to test for acid phosphatase only detect the enzyme if membranes are first disrupted.

   B.  **Acid Phosphatase and Other Acid Hydrolases Are Packed Together in Lysosomes, pp. 287-289**
       1.  In 1955, de Duve introduced the term lysosome for the organelles containing hydrolytic enzymes.
       2.  The lysosomal enzymes are acid hydrolases (with maximal activity at acid pH).

   C.  **Plant Spherosomes are Specialized Lysosomes Exhibiting a High Lipid Content, pp. 289-290**
       1.  Spherosomes are refractive spherical particles visible in the light microscope.
       2.  Spherosomes contain lysosomal enzymes and store lipid reserves.

   D.  **Lysosomes Are Formed by a Pathway That Uses Mannose 6-Phosphate as a Targeting Signal, p. 290**
       1.  Lysosomal enzymes are synthesized on the rough ER and pass through transitional vesicles to the *cis* Golgi network where the N-linked oligosaccharide is converted to mannose 6-phosphate.
       2.  Lysosomal enzymes are packaged into lysosomes at the *trans* Golgi network.
       3.  The pH of the vesicles is lowered to 5 by an ATP-dependent proton pump which releases the enzymes from the membrane and the mannose is dephosphorylated. The mannose 6-phosphate receptors are returned to the Golgi complex.

   E.  **Some Proteins Are Targeted to the Lysosome by a Mannose 6-Phosphate Independent Pathway, pp. 291-292**
       1.  Patients with I-cell disease do not phosphorylate mannose so enzymes are secreted instead of packaged into into lysosomes.
       2.  Lysosomal proteins that are synthesized as membrane bound proteins are not signaled with mannose 6-phosphate; the targeting signal is an amino acid sequence on the cytosolic surface of the membrane.

F. **Lysosomes Are Involved in Macroautophagy, Microautophagy, Autolysis, and Extracellular Digestion, pp. 292-293**
 1. Organelles wrapped in smooth ER fuse with a primary lysosome and are digested in macroautophagy.
 2. Invagination of the lysosomal membrane takes in soluble proteins for digestion in microautophagy.
 3. Autolysis is the digestion of an entire cell.

G. **During Phagocytosis, Plasma Membrane Vesicles Containing Particulate Matter Are Brought into the Cell and Fuse with Lysosomes, pp. 293-294**
 1. Endocytosis is any process that brings materials into a cell in membrane-bound vesicles.
 2. Phagocytosis is endocytosis of particulate matter.
 3. Foreign particles bind to a receptor, the plasma membrane engulfs the particle in a phagosome. The phagosome fuses with a lysosome to form a secondary lysosome.

H. **During Pinocytosis, Plasma Membrane Vesicles Containing Fluid Are Brought into the Cell and Fuse with Lysosomes, pp. 294-295**
 1. In pinocytosis, the plasma membrane randomly folds in to enclose molecules in a pinocytic vesicle.
 2. Lysosomes fuse with pinocytic vesicles to form secondary lysosomes.
 3. Undigested material accumulates in secondary lysosomes to form residual bodies. Residual bodies fuse to the plasma membrane to expel their contents by exocytosis.

I. **Receptor-Mediated Endocytosis Brings Specific Macromolecules into the Cell, pp. 295-297**
 1. A ligand binds to a specific receptor and the receptor-ligand complexes are concentrated on the plasma membrane at coated pits.
 2. The "coating" is a lattice of clathrin molecules organized into a polyhedral "cage."
 3. The coated pit internalizes as a coated vesicle and the vesicle contents accumulate in the early endosome.
 4. Dissociation of the ligand and receptor occurs in the late endosome or CURL.
 5. The receptor can be returned to the plasma membrane while the ligand goes to a lysosome.
 6. The receptor and ligand can be returned to the membrane by carrier vesicles.
 7. The receptor and ligand can go to a lysosome.
 8. The receptor and ligand can be delivered to the opposite side of the cell to fuse with the plasma membrane; a process called transcytosis.
 9. Caveolae are invaginations of the plasma membrane containing receptors or enzymes. Small molecules or ions that bind are driven into the cell by diffusion (potocytosis).

J. **Receptor-Mediated Endocytosis Is Utilized for Several Purposes, Including Nutrient Delivery and Signal Transduction, pp. 297-298**
 1. An LDL receptor binds LDL, the LDL receptor is recycled to the plasma membrane and the apo-B protein is degraded and the cholesterol and phospholipids are released for use within the cell.
 2. After a signal (e.g., hormone or growth factor) binds to a receptor, the ligand-receptor is taken in by endocytosis and degraded thereby decreasing the number of receptors (down-regulation).

Chapter 7

    3. Transcytosis is used to move proteins from one side of a cell layer to the other.

  K. **Lysosomal Storage Diseases Are Caused by Genetic Defects in Lysosomal Enzymes, pp. 298-300**
    1. Lysosomal storage diseases are characterized by excessive intracellular accumulation of polysaccharides or lipids.
    2. β-glycosidase is lacking in type II glycogenosis (abnormal storage of glycogen).
    3. Missing enzymes can be incorporated into liposomes for treatment.

VI. **MEMBRANE BIOGENESIS, pp. 300-305**
    1. Lipids and proteins can be inserted into preexisting membranes.

  A. **How Are Phospholipids Inserted into Cellular Membranes? pp. 302-303**
    1. Membranes are synthesized in the ER. Phospholipid translocators (flippases) are responsible for the asymmetry of the membrane.
    2. Phospholipid transfer proteins move phospholipids from the ER to other membranes.
    3. Bacterial phospholipids are synthesized in the cytosol side of the plasma membrane and can be moved to the outer leaflet.
    4. Membrane asymmetry may be due to either of the following:
      a. Phospholipid translocators are substrate specific.
      b. Some phospholipids may be thermodynamically more stable on one side of the membrane.

  B. **How Are Integral and Peripheral Membrane Proteins Incorporated into Cellular Membranes? pp. 303-304**
    1. Membrane proteins (e.g., VSV M protein) are synthesized on free ribosomes and released into the cytosol.
    2. Membrane proteins (e.g., VSV glycoprotein) contain a signal sequence for synthesis on the ER and fusion with the plasma membrane.

  C. **How Are Lipid-Anchored Proteins Incorporated into Cellular Membranes? pp. 304-305**
    1. Glycolipid-anchored membrane proteins are bound to GPI. These proteins are synthesized on the rough ER, fused to the inside of vesicles in the Golgi complex, and face the exterior of the cell when the vesicle fuses with the plasma membrane.
    2. Fatty acid-anchored membrane proteins are bound to the cytoplasmic surface of the membrane by myristic acid or palmitic acid. They are synthesized as soluble cytosol proteins.
    3. Prenylated membrane proteins are bound to isoprenoid lipid groups (prenylation). These proteins are synthesized and prenylated in the cytosol.

# KEY TERMS

| | | |
|---|---|---|
| acid hydrolase | annexin | brefelden A |
| bulk flow | caveolae | chaperone |
| *cis* face | cisternae | cisternal space |
| clathrin | clathrin | coated pits |
| coated vesicle | condensing vesicle | constitutive secretion |
| cytochrome P-448 | cytochrome P-450 | cytosol |

# Chapter 7

| | | |
|---|---|---|
| endocytosis | endoplasmic reticulum | ER lumen |
| exocytosis | flippases | glycoprotein |
| Golgi complex | GPI | liposome |
| lysosomal storage disease | lysosome | macroautophagy |
| mannose 6-phosphate | microautophagy | microsome |
| mixed-function oxidase | N-linked glycosylation | NSF |
| O-linked glycosylation | phagocytosis | phagosome |
| phospholipid transfer protein | phospholipid translocators | pinocytosis |
| potocytosis | prenylation | primary lysosome |
| proenzyme | prohormone | protein disulfide isomerase |
| protein-translocating channel | receptor-mediated endocytosis | regulated secretion |
| residual body | rough ER | secondary lysosome |
| secretory vesicle | signal recognition particle | signal peptidase |
| signal sequence | smooth ER | SNAP |
| SNARE | stop-transfer sequence | targeting signals |
| transcytosis | *trans* face | transport vesicle |

## ⇗ KEY FIGURE

Label the following structures: nucleus, smooth ER, rough ER, cisternal space, cytosol, and Golgi complex. Show how secreted proteins move from the ER to the environment. Show how an enzyme moves from the ER to a lysosome, and to a phagosome.

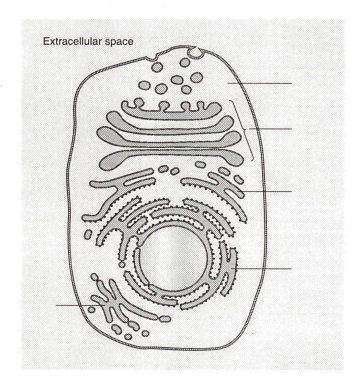

# STUDY QUESTIONS

1. Which of the following is not associated with the ER?
   a. Carbohydrate metabolism
   b. Membrane synthesis
   c. Oxidation of foreign substances
   d. Protein processing
   e. None of the above

2. The following steps are involved with protein synthesis on the rough ER. What is the third step?
   a. The polypeptide chain attaches to the protein-translocating channel
   b. The ribosome dissociates
   c. The signal sequence is removed by signal peptidase
   d. The SRP binds to SRP receptor
   e. The SRP binds to the signal sequence

3. Integral membrane proteins are inserted into the membrane with the N-terminal end facing the lumen by a
   a. Targeting signal.
   b. Signal sequence.
   c. Signal recognition particle.
   d. Stop-transfer sequence.
   e. Chaperone.

4. Which of the following enzymes causes dissociation of the ribosome from the ER?
   a. Glycosylase
   b. Signal peptidase
   c. Flippase
   d. Phospholipid translocators
   e. None of the above

5. Which one of the following is **not** a function of the Golgi complex?
   a. Formation of lysosomes
   b. Synthesis of lysosomal enzymes
   c. Formation of the plasma membrane
   d. Synthesis of hyaluronate and pectin
   e. None of the above

6. Which one of the following is **not** true of material entering through a coated pit?
   a. Receptor and ligand return to the plasma membrane
   b. Receptor and ligand are degraded by a lysosome
   c. Receptor and ligand move to the opposite side of the cell
   d. Receptor and ligand are stored in vesicles
   e. None of the above

Use this diagram to answer questions 7-10.

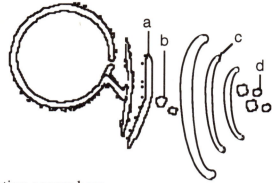

7. O-linked glycosylation occurs here.

8. Structure **d** could become any of the following **except** a
   a. Storage vesicle.
   b. Lysosome.
   c. Part of the plasma membrane.
   d. Mitochondrion.
   e. None of the above

9. Polypeptides get from **a** to **c** because of
   a. An N-terminal amino acid sequence.
   b. An N-linked oligosaccharide.
   c. An O-linked oligosaccharide.
   d. A ribosome.
   e. None of the above

10. Proinsulin is converted to active insulin here.

11. Mannose 6-phosphate targets proteins for incorporation into
    a. Lysosomes.
    b. Golgi complex.
    c. The plasma membrane.
    d. Mitochondria.
    e. Secretory vesicles.

12. Ethyl alcohol is detoxified in the liver. You would expect alcohol to have which of the following effects on liver cells?
    a. Nuclear degeneration
    b. Growth of the smooth ER
    c. Increased lysosomes
    d. Growth of rough ER
    e. None of the above

13. Many bacterial pathogens require iron to grow. Human cells sequester the iron using
    a. Phagocytosis.
    b. Receptor-mediated endocytosis.
    c. Transitional vesicles.
    d. Secretory vesicles.
    e. None of the above

# Chapter 7

14. Which of the following is not correctly matched?
    a. Phagocytosis—particles
    b. Pinocytosis—soluble macromolecules
    c. Potocytosis—small molecules
    d. Coated pits—insulin
    e. None of the above

15. Pancreatic cells secrete trypsinogen when stimulated by a hormone called CCK. This is an example of
    a. Constitutive secretion.
    b. Bulk flow.
    c. Regulated secretion.
    d. Transcytosis.
    e. None of the above

16. Which of the following first binds to the ER for protein synthesis?
    a. Growing polypeptide
    b. mRNA
    c. Ribosome subunits
    d. Signal sequence
    e. SRP

17. Which of the following is **not** associated with the Golgi complex?
    a. Glucan synthetases
    b. Glycosyl transferases
    c. Nucleoside diphosphatase
    d. Signal peptidase
    e. Thiamine pyrophosphatase

18. Studies using tunicamycin, which blocks N-linked glycosylation, have shown that some proteins made in the ER are secreted. This indicates that
    a. A chemical signal is needed for protein synthesis.
    b. No chemical signal is needed for transitional vesicles.
    c. The smooth ER is involved in some protein synthesis.
    d. Secretion is constitutive.
    e. None of the above

In a pulse-chase experiment, where would you find the materials listed in questions 19-21?
    a. In the ER
    b. In the plasma membrane
    c. Outside the cell
    d. Secretory vesicles
    e. In the Golgi complex

19. A $^3$H-mannose.

20. The radioactive label in cells whose secretion has been inhibited.

21. Clathrin triskelions.

# PROBLEMS

1. Congenital sucrase-isomaltase deficiency (CSID) results in osmotic diarrhea because the patient cannot metabolize sucrose. CSID patients make sucrase-isomaltase and permanently store it in the ER. Provide a possible mechanism for this disease.

2. Hunter's disease and Hurler's disease are inherited conditions in which patients accumulate chondroitin in their cells. If Hunter's fibroblasts and Hurler's fibroblasts are combined *in vitro* they will cure each other. Explain these diseases.

3. Surface proteins of cultured mouse cells were labeled with $^{125}$I using lactoperoxidase and treated with antibodies to the insulin receptor to locate the insulin receptor. A different cell culture was labeled and incubated at 37°C for 1 hour prior to treatment with antibodies to the insulin receptor. The experiment was repeated keeping the cells at 10°C for 1 hour. What happened to the labeled receptors at 37°C?

|  | Radioactive proteins bound to antibodies |
|---|---|
| No incubation | 100% (control) |
| 37°C | 65% |
| 10°C | 100% |

4. In cells infected with *Varicellavirus*, the viral DNA and protein coat are synthesized in the nucleus. Pulse-chase experiments using $^3$H-fucose have shown the viral glycoproteins are bound to the plasma membrane 4 hours after infection. Describe how the viral glycoprotein is made in the cell.

5. In plant cells, secretory vesicles formed in the Golgi complex fuse with the plasma membrane and deliver cell wall materials to the exterior of the cell. During growth, enough new membrane is added to double the size of the plasma membrane every 20 minutes. How does the plant cell limit the expansion of the membrane?

6. The number of chaperone proteins increases in cells in the presence of amino acid analogs and puromycin. Suggest a reason for this.

7. Assume that you genetically engineer a hamster cell culture to produce an enzyme. You would like the enzyme secreted from the cells for easy collection. What genes do you need in addition to the gene coding for the desired enzyme?

# ANSWERS TO STUDY QUESTIONS

| 1. e | 2. a | 3. d | 4. b |
| 5. b | 6. d | 7. c | 8. d |
| 9. a | 10. d | 11. a | 12. b |
| 13. b | 14. e | 15. c | 16. e |
| 17. d | 18. b | 19. e | 20. d |
| 21. b | | | |

# CHAPTER 8

## *Mitochondria and the Capturing of Energy Derived from Food*

## 🏠 LEARNING OBJECTIVES

Be able to
1. Define oxidation-reduction reaction.
2. Explain the role of NADH in oxidative phosphorylation.
3. Define each of the following and identify the purpose of each process:
   a. Glycolysis
   b. Fermentation
   c. Krebs cycle
   d. Respiratory electron transport chain
4. Compare and contrast
   a. Aerobic and anaerobic metabolism
   b. Glycolysis and fermentation
   c. Glycolysis and the Krebs cycle
   d. Eukaryotic electron transport and prokaryotic electron transport
   e. Oxidative phosphorylation and substrate-level phosphorylation
5. Show where each of the following occur in a mitochondrion:
   a. β-oxidation pathway
   b. Krebs cycle
   c. ATP formation
6. Describe the respiratory electron transport chain. Identify the coupling sites and how ATP synthase works.
7. Provide the function of peroxisomes, glyoxysomes, and glycosomes.

## 🏠 CHAPTER OVERVIEW

**I.   INTRODUCTION, p. 310**
   1. In anaerobic metabolism, a small percentage of the energy stored in food is captured in metabolic processes; anaerobic metabolism occurs in the cytosol.
   2. In aerobic metabolism, energy stored in food is captured in metabolic processes that occur in eukaryotic mitochondria or prokaryotic plasma membranes.

**II.  ANAEROBIC PATHWAYS FOR CAPTURING ENERGY, pp. 310-315**

   **A.  Energy Is Released by Oxidation Reactions, pp. 310-311**
   1. Oxidation is the loss of electrons; this is exergonic.
   2. Reduction is the gain of electrons.

   **B.  Coenzymes and ATP Play Central Roles in Transferring Energy from One Reaction to Another, p. 311**
   1. Flavoproteins use FAD and FMN as coenzymes. $NAD^+$ and $NADP^+$ are also used as coenzymes.
   2. Each coenzyme can accept a pair of electrons.
   3. When reduced, the coenzymes (NADH, NADPH, $FADH_2$, and $FMNH_2$) carry energy that can be transferred to another reaction.

C. **Glucose Plays a Central Role as a Source of Both Energy and Chemical Building Blocks, p. 311**
1. Other substrates can be oxidized in glycolysis.

D. **Glycolysis is the First Stage in Extracting Energy from Glucose, pp. 311-313**
1. Embden, Meyerhof, and Warburg discovered that glycolysis is a series of 10 reactions.
2. The summary reaction is:
Glucose → 2 Pyruvate + 2 ATP + 2 NADH.

E. **Fermentation Replenishes $NAD^+$ While Reducing Pyruvate to Lactate or Ethanol, p. 313**
1. In alcoholic fermentation, yeast produce $CO_2$ and ethyl alcohol to reoxidize NADH.
Glucose → 2 Ethyl alcohol + 2 $CO_2$ + 2 ATP
2. In lactate fermentation, muscle cells produce lactic acid to reoxidize NADH.
Glucose → 2 Lactic acid + 2 ATP
3. Bacteria can produce a variety of products in fermentation.

F. **The 2 ATP Molecules Produced during the Fermentation of Glucose Represent a Relatively Small Energy Yield, pp. 313-315**
1. Glucose + 6 $O_2$ → 6 $CO_2$ + 6 $H_2O$     $\Delta G^{\circ\prime}$ = -686 kcal/mol
2. Glucose → 2 Lactic acid     $\Delta G^{\circ\prime}$ = -22 kcal/mol
3. Complete oxidation of glucose is accomplished in cellular respiration.
4. Pyruvate is oxidized in the Krebs cycle.
5. Electrons generated in the Krebs cycle are used to generate ATP by electron transport and oxidative phosphorylation.

III. **ANATOMY OF THE MITOCHONDRION, pp. 315-325**

A. **Mitochondria Were First Discovered and Functionally Described Using Microscopy and Subcellular Fractionation, pp. 315-316**
1. Michaelis discovered that mitochondria stain green with Janus green; but became decolorized as the cells consumed oxygen.
2. Kennedy and Lehninger discovered that the Krebs cycle, electron transport, and oxidative phosphorylation occur in mitochondria.

B. **Mitochondria Consist of Outer and Inner Membranes That Define Two Separate Compartments, p. 316**
1. The mitochondrion is surrounded by the outer mitochondrial membrane.
2. The inner mitochondrial membrane is folded into cristae.
3. Cristae are covered with 9-nm particles called $F_1$ particles.
4. The intermembrane space is located between the outer and inner membranes and the matrix is enclosed by the inner mitochondrial membrane.

C. **Mitochondria Can Form Large Interconnected Networks, pp. 316-319**
1. Mitochondria are located near energy-requiring structures or energy-rich substrates.
2. Mitochondria form interconnected networks that pinch off or fuse with one another.

## Chapter 8

- D. **Mitochondrial Membranes and Compartments Can Be Separated from Each Other for Biochemical Study, p. 319**
    1. In a hypotonic solution containing digitonin, the outer membrane ruptures releasing the contents of the intermembrane space. The outer and inner membranes can be separated by centrifugation.
    2. Inner membranes form vesicles called mitoplasts. The membrane and matrix can be separated by disrupting the inner membrane with Lubrol.
    3. The inner membrane reseals itself to form inner membrane vesicles with $F_1$ particles protruding to the exterior.

- E. **The Outer Mitochondrial Membrane is Permeable to Small Molecules, pp. 320-322**
    1. The outer membrane is involved in the degradation of lipids and synthesis of membrane phospholipids.
    2. The outer membrane contains monoamine oxidase which breaks down norepinephrine and dopamine.
    3. The outer membrane contains porins which permit the passage of smaller molecules into the intermembrane space.
    4. Adenylate kinase (which catalyzes AMP + ATP $\leftrightarrow$ 2 ADP) occurs on the outer membrane.

- F. **The Mitochondrial Matrix Is the Site Where the Krebs Cycle Occurs, pp. 322-323**
    1. The matrix contains Krebs cycle enzymes, DNA, RNA, ribosomes, and enzymes involved in nucleic acid and protein synthesis.

- G. **The Inner Mitochondrial Membrane Is the Main Site of Mitochondrial ATP Formation, pp. 323-324**
    1. The inner membrane is high in cardiolipid and unsaturated phospholipids and has little cholesterol.
    2. The inner membrane is impermeable to most nucleotides, sugars, and small ions.
    3. Steroid hormones are synthesized on the inner membrane.

IV. **HOW MITOCHONDRIA CAPTURE ENERGY, pp. 324-344**

- A. **Pyruvate and Fatty Acids are Oxidized to Acetyl CoA in the Mitochondrial Matrix, pp. 324-326**
    1. Pyruvate dehydrogenase catalyzes: Pyruvate + $NAD^+$ $\rightarrow$ Acetyl CoA + $CO_2$ + NADH.
    2. Fats are hydrolyzed in the cytosol to fatty acids.
    3. Fatty acids are hydrolyzed by β-oxidation in the matrix to acetyl CoA.
    4. β-oxidation produces 1 NADH and 1 $FADH_2$ for every acetyl CoA.

- B. **The Oxidation of Acetyl CoA in the Krebs Cycle Generates a Small Amount of ATP Plus NADH and $FADH_2$, p. 326**
    1. Acetyl CoA combines with oxaloacetate to produce citrate.
    2. Citrate is oxidized in the Krebs cycle to produce 3 NADH, 1 $FADH_2$, and 1 ATP; 2 $CO_2$ are released.
    3. Krebs cycle enzymes exist in the mitochondrial matrix, except succinate dehydrogenase which is an integral protein in the inner mitochondrial membrane.

# Chapter 8

C. **NADH and $FADH_2$ are Oxidized Back to $NAD^+$ and FAD by the Respiratory Chain, pp. 326-327**
  1. The respiratory chain is a series of electron carriers to reoxidize NADH and $FADH_2$:
     $NADH + {}^1/_2 O_2 \rightarrow NAD^+ + H_2O$
     $FADH_2 + {}^1/_2 O_2 \rightarrow FAD + H_2O$
  2. The flow of electrons through the respiratory chain is called electron transport and the accompanying formation of ATP is oxidative phosphorylation.

D. **The Respiratory Chain Is Comprised of Four Different Classes of Molecules, pp. 327-328**
  1. **Cytochromes consists of a heme (porphyrin ring bound to an iron atom), p. 327-328**
     a. The iron is reduced by the transfer of an electron: $Fe^{2+} \rightarrow Fe^{3+}$.
     b. At least 5 cytochromes are present in the respiratory chain: cytochromes $b$, $c$, $c_1$, $a$, and $a_3$.
     c. Cytochromes (except $c$) are integral membrane proteins in the inner membrane.
     d. Cytochromes $a$ and $a_3$ contain Cu instead of Fe.
  2. **Iron-sulfur centers transfer electrons between $Fe^{3+}$ and $Fe^{2+}$, p. 328**
  3. **Flavoproteins are enzymes that contain FMN or FAD, p. 328**
  4. **Ubiquinones or coenzyme Q are lipids in the inner membrane that can accept two electrons, p. 328**

E. **Redox Potentials Indicate the Energetically Most Favorable Route for Electrons Passing Through the Respiratory Chain, pp. 328-330**
  1. A pair of ions consisting of the oxidized and reduced forms is a redox couple.
  2. $-E'_0$ means the redox couple has a tendency to donate electrons (to a redox couple that is less negative).
  3. $\Delta G^{\circ\prime} = -nF\Delta E'_0$; where n = number of electrons and F = 23,062.

F. **Difference Spectra Reveal the Sequence of the Respiratory Chain within Intact Mitochondria, pp. 330-331**
  1. The absorption spectra of respiratory carriers can be used to calculate the relative degree of oxidation of each carrier.
  2. The eukaryotic respiratory chain is: $NAD^+ \rightarrow FMN \rightarrow$ Iron-sulfur center $\rightarrow$ Ubiquinone $\rightarrow$ Cytochrome $b \rightarrow$ Iron-sulfur center $\rightarrow$ Cytochrome $c_1 \rightarrow$ Cytochrome $c \rightarrow$ Cytochrome $a \rightarrow$ Cytochrome $a_3 \rightarrow {}^1/_2 O_2$.

G. **The Respiratory Chain Is Organized into Four Multiprotein Complexes, pp. 331-332**
  1. Respiratory complex I transfers electrons from NADH to ubiquinone.
  2. Respiratory complex II transfers electrons from succinate to ubiquinone.
  3. Respiratory complex III transfers electrons from ubiquinone to cytochrome $c$.
  4. Respiratory complex IV transfers electrons from cytochrome $c$ to $O_2$.
  5. Carriers diffuse freely through the inner membrane.

H. **The Respiratory Chain Contains Three Coupling Sites for Oxidative Phosphorylation, pp. 332-333**
  1. 3 ATPs are produced for each NADH oxidized and 2 ATPs from each $FADH_2$ oxidized.

# Chapter 8

    2. Complexes I, III, and IV are coupled to oxidative phosphorylation.

**I. The $F_1$-$F_0$ Complex Is the Site of ATP Synthesis during Oxidative Phosphorylation, pp. 333-335**
1. $F_1$ protrudes from the surface of the inner membrane and is attached to $F_0$ which spans the inner membrane.
2. The $F_1$-$F_0$ Complex is called ATP synthase.

**J. An Electrochemical Proton Gradient Across the Inner Mitochondrial Membrane Drives ATP Formation, pp. 335-340**
1. In 1962, Mitchell proposed a model called chemiosmotic coupling.
2. Energy released during electron transfer is used to pump $H^+$ from the matrix into the intermembrane space creating an electrochemical proton gradient consisting of $\Delta\psi$ and $\Delta pH$; where $\Delta\psi$ is the membrane potential.
3. ATP is synthesized as the protons move down their electrochemical gradient.
4. The electrochemical gradient exerts a proton motive force which drives protons down their concentration gradient.
5. Uncoupling agents abolish both the proton gradient and oxidative phosphorylation.
6. Oxidative phosphorylation requires an intact membrane defining an enclosed compartment.
7. The components of the respiratory chain are asymmetrically oriented within the inner membrane.
8. ATP is synthesized as protons flow through the $F_1$-$F_0$ Complex.

**K. A Maximum of 38 Molecules of ATP Is Produced per Molecule of Glucose Oxidized, pp. 340-342**
1. The net equation for aerobic glucose oxidation is
$C_6H_{12}O_6 + 6\,O_2 \rightarrow 6\,CO_2 + 6\,H_2O$.
2. In eukaryotic cells, NADH made in the cytosol transfers electrons to carriers which in turn transfer the electrons to the electron carriers in the mitochondria.
3. The malate-aspartate shuttle transfers electrons to $NAD^+$.
4. The glycerol phosphate shuttle transfers electrons to FAD.
5. ATP made in glycolysis is made by substrate-level phosphorylation.
6. Of the energy stored in a glucose molecule, 60% is captured in ATP and 40% is lost as heat.

**L. Molecules Involved in Mitochondrial Metabolism Are Actively Transported Across the Inner Mitochondrial Membrane, pp. 342-343**
1. $H^+$-linked carriers transport pyruvate and inorganic phosphate into the matrix.
2. ADP is brought in by an ADP-ATP antiport.
3. Transport across the inner membrane uses some of the energy of the proton gradient.

**M. Allosteric Regulation Plays an Important Role in Controlling Glucose Oxidation, pp. 343-344**
1. The Pasteur effect says that glucose consumption decreases in the presence of oxygen.
2. The presence of $O_2$ inhibits step 3 of glycolysis.
3. ATP is an allosteric inhibitor of the glycolytic enzyme phosphofructokinase; AMP is an allosteric activator of this enzyme.

# Chapter 8

       4. ATP is an allosteric inhibitor of pyruvate dehydrogenase; AMP is an allosteric activator of this enzyme.
       5. ADP is an allosteric activator of the Krebs cycle enzyme isocitrate dehydrogenase.

  **N. Respiratory Control Regulates the Flow of Electrons through the Respiratory Chain, p. 344**
       1. Respiratory control ensures that electron transfer occurs only when ATP can be made.
       2. If ADP is not available, the proton gradient becomes so steep that no more protons can be pumped.
       3. Cells producing large amounts of ATP have increased cristae (condensed configuration) while cells not producing ATP have few cristae (orthodox configuration).

## V. HOW BACTERIA CAPTURE ENERGY, pp. 344-347

  **A. The Respiratory Chain and Oxidative Phosphorylation Occur in the Plasma Membrane of Bacterial Cells, pp. 344-346**
       1. Prokaryotic cells use the same electron carriers but show variation in the sequence.
       2. $H^+$ pumped out of the bacterial cell by the electron transfer chain.

  **B. An Electrochemical Proton Gradient Across the Bacterial Plasma Membrane Drives Oxidative Phosphorylation, pp. 346-347**
       1. The proton gradient is used to generate ATP and to facilitate active transport (e.g., of lactose-$H^+$ symport).
       2. *E. coli* produces antibiotics, called colicins E1 and K, which cause $K^+$ to leak into cells thus disrupting the electrochemical gradient.

## VI. PEROXISOMES AND RELATED ORGANELLES, pp. 347-352
       1. Peroxisomes are involved in different functions in different cells.

  **A. Peroxisomes Were Discovered by Isodensity Centrifugation, pp. 347-348**
       1. Peroxisomes are distinguished from other microbodies by a crystalline core containing urate oxidase and the presence of catalase.

  **B. Peroxisomes Contain Enzymes Involved in the Production and Breakdown of Hydrogen Peroxide, pp. 348-349**
       1. Urate oxidase, amino acid oxidase, and glycolate oxidase generate $H_2O_2$ by: $RH_2 + O_2 \rightarrow R + H_2O_2$.
       2. In the catalatic mode:
$$2\,H_2O_2 \xrightarrow{catalase} O_2 + 2\,H_2O$$
       3. In the peroxidatic mode:
$$RH_2 + H_2O_2 \xrightarrow{catalase} R + 2\,H_2O$$

  **C. Peroxisomes Perform Several Metabolic Functions, pp. 349-350**
       1. Inactivation of toxic substances occurs by the peroxidatic mode.
       2. Peroxisomes can oxidize substances using molecular oxygen thereby protecting the cell from high concentrations of oxygen.
       3. Lipid metabolism occurs in peroxisomes.

4. The presence of urate oxidase indicates that peroxisomes have a role in degradation of purines.
5. Peroxisomes may be involved in synthesizing carbohydrates from lipids (gluconeogenesis).

D. **Glyoxysomes Are Used by Plants to Synthesize Carbohydrates from Lipids, pp. 350-351**
1. In plant cells, stored lipids are converted to acetyl CoA by β-oxidation.
2. The acetyl CoA is then converted to carbohydrates in the glyoxylate cycle.
3. Enzymes of the glyoxylate cycle, β-oxidation enzymes, catalase, and hydrogen peroxide-producing oxidases are located in glyoxysomes.

E. **Glycosomes Are Used by Trypanosomes to Speed Up Glycolysis, pp. 351-352**
1. Glycosomes occur in trypanosomes.
2. Glycosomes contain the first 7 glycolytic enzymes, catalase, and β-oxidation enzymes.

# KEY TERMS

| | | |
|---|---|---|
| aerobic | anaerobic | ATP synthase |
| β-oxidation pathway | cellular respiration | chemiosmotic coupling |
| coenzyme Q | coupling site | coupling site |
| cristae | cytochrome | difference spectrum |
| electrochemical proton gradient | electron transport | $F_1$ particles |
| F1-F0 complex | flavoprotein | gluconeogenesis |
| glycerol phosphate shuttle | glycolysis | glycosome |
| glyoxylate cycle | glyoxysome | heme |
| inner mitochondrial membrane | intermembrane space | Krebs cycle |
| malate-aspartate shuttle | matrix | outer mitochondrial membrane |
| oxidation | oxidative phosphorylation | Pasteur effect |
| peroxisome | proton motive force | redox couple |
| reduction | respiratory chain | respiratory control |
| standard redox potential ($E'_0$) | substrate-level phosphorylation | ubiquinone |
| uncoupling agent | | |

# Chapter 8

## KEY FIGURE

Label the crista, matrix, and intermembrane space. Label each of the respiratory complexes, ubiquinone, FAD, cytochrome *c*, cytochrome oxidase, and ATP synthase. Diagram the path of electrons and protons. Which side of the membrane is acidic? Show where ATP is made.

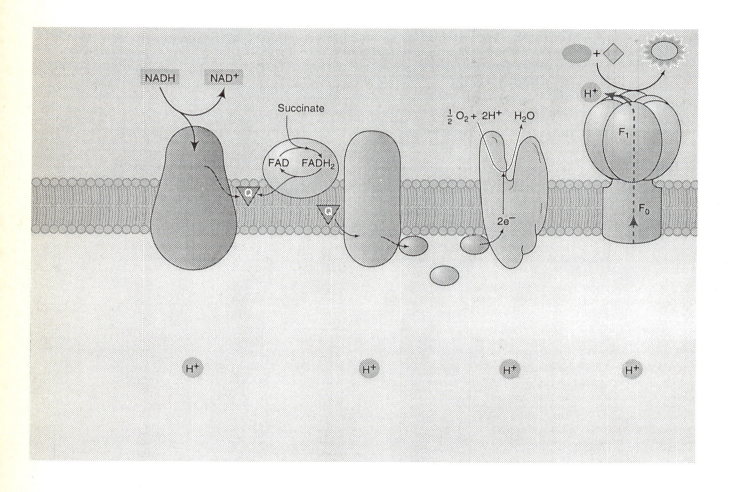

# STUDY QUESTIONS

1. In a eukaryotic cell, what percentage of the ATPs made from the complete oxidation of glucose are produced by oxidative phosphorylation?
    a. 100%
    b. 94%
    c. 88%
    d. 12%
    e. 0%

2. Which of the following is **not** an end product of fermentation?
    a. Acetone
    b. ATP
    c. Butyric acid
    d. Ethyl alcohol
    e. Lactic acid

Use this figure to answer questions 3-6.

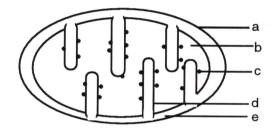

3. The location of the Krebs cycle.

4. Location of the respiratory electron carriers.

5. The most acidic area.

6. In the figure, c is the site of
    a. Proton pumps.
    b. Phosphorylation of ADP.
    c. Cytochromes.
    d. The Krebs cycle.
    e. Electrons.

7. The antibiotic antimycin blocks the flow of electrons between cytochromes *b* and *c*. Why isn't antimycin used to treat bacterial diseases?
    a. Bacteria don't have mitochondria.
    b. Bacteria don't have cytochromes *b* and *c*.
    c. Antimycin is toxic to human cells.
    d. Cells would switch to glycolysis.
    e. None of the above.

8. How many molecules of carbon dioxide will be given off if you supply an aerobic cell with 5 pyruvic acid molecules?
    a. 5
    b. 10
    c. 15
    d. 20
    e. 34

9. For a cell, which of the following compounds has the greatest amount of energy per molecule?
    a. ATP
    b. ADP
    c. NADH
    d. $O_2$
    e. Acetyl CoA

The stages of glucose oxidation in a eukaryotic cell are listed below. Use these choices to answer questions 10 and 11. Choices may be used once, more than once, or not at all.
    a  Glycolysis
    b. Production of acetyl CoA
    c. Krebs cycle
    d. Oxidative phosphorylation

10. Which stage produces the most ATP?

11. Which stage occurs in the presence or absence of $O_2$?

12. 500 molecules of $^{14}C$-glucose are given to yeast cells that are producing alcohol. How many molecules of $^{14}C$-ATP molecules can be made from this glucose?
    a. 0
    b. 2
    c. 4
    d. 400
    e. 2000

13. Your biotechnology company is growing a large volume of cells but they are not producing the desired product because the medium is too acidic. What is probably wrong?
    a. The cells are producing too much ATP.
    b. Glycolysis is not occurring.
    c. There is too little $O_2$ in the medium.
    d. The cells using the Krebs cycle.
    e. There is too much sugar in the medium.

14. Which of the following reactions yields the most new ATP molecules under aerobic conditions?
    a. Acetyl CoA $\rightarrow$ $CO_2$ + $H_2O$
    b. Pyruvic acid $\rightarrow$ acetyl CoA
    c. Glucose $\rightarrow$ pyruvic acid
    d. Phosphoglyceraldhyde $\rightarrow$ 1 glucose
    e. NADH + $H^+$ $\rightarrow$ $NAD^+$ + $H_2O$

# Chapter 8

15. What source of energy is used to move protons in chemiosmosis?
    a. Protons push each other along
    b. Active transport
    c. Oxidation of carriers
    d. Energy from ATP
    e. No energy is required since the process is exergonic.

16. Red blood cells lack mitochondria. When a red blood cell is metabolizing glucose all of the following substances are produced **except**
    a. Lactic acid.
    b. Phosphoglyceraldehyde.
    c. Acetyl CoA.
    d. Pyruvic acid.
    e. ATP.

17. The enzymes of glycolysis are located
    a. On the inner surface of the plasma membrane.
    b. In the cytosol.
    c. In the inner membrane of the mitochondrion.
    d. In the peroxisomes.
    e. In the Golgi apparatus.

18. The presence of oxygen inhibits glycolysis at phosphofructokinase. The mechanism of inhibition is
    a. Allosteric inhibition by AMP.
    b. Feedback inhibition by pyruvic acid.
    c. Competitive inhibition by lactic acid.
    d. Allosteric inhibition by ATP.
    e. Lack of fructose.

19. When an animal cell is metabolizing glucose in the complete absence of molecular oxygen, which one of the following substances is not produced?
    a. Acetyl CoA
    b. ATP
    c. Lactic acid
    d. NADH
    e. Pyruvic acid

20. The purpose of fermentation is to
    a. Prevent oxidative phosphorylation.
    b. Produce alcohol.
    c. Oxidize NADH.
    d. Produce glucose.
    e. Produce pyruvate.

21. How much molecular oxygen is required in the fermentation of one molecule of glucose?
    a. None
    b. 1 molecule
    c. 24 molecules
    d. 36 molecules
    e. 38 molecules

# Chapter 8

## 👉 PROBLEMS

1. Animals living in cold places have brown fat. Brown fat gets its color from the presence of cytochromes. In brown fat, electron transport is uncoupled from ATP synthesis.
   a. What is the advantage of brown fat?
   b. What is the advantage of the uncoupling?

2. Why can prokaryotes make 38 ATPs from 1 molecule of glucose and eukaryotes only make 34 ATPs?

3. Bacteria can oxidize NADH by the reduction of pyruvate, $O_2$, nitrate ions, or methylene blue. Using the following standard redox potentials, calculate the free energy for transferring 2 electrons to each of these substrates. If there are no other variables, which electron acceptor is preferred? Briefly explain why.

   | Oxidized form | Reduced form | $E'_0$ (volts) |
   |---|---|---|
   | Pyruvate | Lactate | -0.19 |
   | NAD | NADH | -0.32 |
   | $1/2\ O_2 + 2H^+$ | $H_2O$ | +0.82 |
   | $NO_3^- + 2H^+$ | $NO_2^-$ | +0.42 |
   | Methylene blue (oxidized) | Methylene blue (reduced) | 0.01 |

4. Patients with chronic granulomatous disease suffer from frequent infections because their white blood cells cannot make $H_2O_2$ to kill bacteria. Suggest possible sites for this defect in the cells.

5. Patients with Zellweger Syndrome die within a few months after birth due to the accumulation of fatty acids in cells. What organelle is missing from these patients?

6. *Streptococcus* and *Clostridium* bacteria lack catalase. Why are *Clostridium* bacteria killed by $O_2$ while the streptococci can grow in the presence of $O_2$?

7. Use the diagram on the next page to answer the following questions:
   a. Metabolism of citric acid ($C_6H_8O_7$) occurs here.
   b. What steps require a molecule of ATP?
   c. At what steps is $CO_2$ released?
   e. What is intermediate #9?
   f. At what steps is ATP made by substrate-level phosphorylation?
   g. Hydrocarbons are hydrolyzed by beta-oxidation and catabolized at what intermediates?
   h. If the amino acid, aspartic acid, is decarboxylated and deaminated the remaining residue would most likely be metabolized at what intermediate?

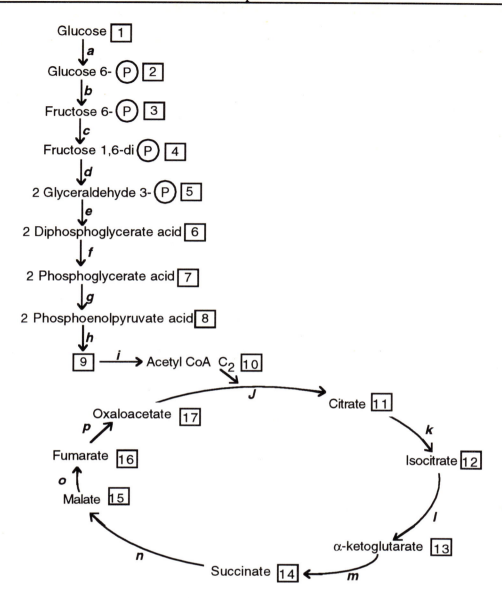

## ☞ ANSWERS TO STUDY QUESTIONS

1. c   2. b   3. b   4. d
5. e   6. b   7. c   8. c
9. e   10. d   11. a   12. a
13. c   14. a   15. c   16. c
17. b   18. d   19. a   20. c
21. a

# CHAPTER 9

# *Chloroplasts and the Capturing of Energy Derived from Sunlight*

## LEARNING OBJECTIVES

Be able to
1. Define the following terms and provide a function for each:
    a. Photosynthesis
    b. $CO_2$ fixation
    c. Pyrenoid
    d. Photosystem
    e. Photosynthetic electron transfer chain
2. Compare and contrast the following pairs of terms:
    a. Autotroph and heterotroph
    b. Light reaction and dark reaction
    c. The anatomy of a mitochondrion and a chloroplast
    d. Chlorophyll *a* and chlorophyll *b*
    e. Chlorophyll and accessory pigments
    f. Light-harvesting complex and reaction-center complex
    g. Photosystem I and photosystem II
    h. Cyclic and noncyclic photophosphorylation
    i. Photophosphorylation and oxidative phosphorylation
    j. Prokaryotic and eukaryotic photosynthesis
3. Describe the contributions made by each of the following: Van Helmont, Priestly, Ingenhousz, Senebrier, Mayer, Blackman, Hill, and Calvin.
4. Describe the value of accessory pigments.
5. Describe the photosynthetic electron transfer chain. Identify the coupling sites and how ATP synthase works.
6. List four factors that regulate rubisco activity. Describe how the presence of $O_2$ affects rubisco.
7. Compare and contrast $C_3$ plants with $C_4$ plants and CAM plants.
8. Describe the generation of ATP by bacteriorhodopsin.

## CHAPTER OVERVIEW

I.  **INTRODUCTION, p. 355**
    1.  Photosynthesis uses energy from sunlight to reduce carbon dioxide; oxygen and carbohydrates are produced.
    2.  Autotrophs use photosynthesis; these include plants, algae, and the prokaryotic photosynthetic bacteria and cyanobacteria.
    3.  Heterotrophs use organic carbon sources.
    4.  Photosynthesis occurs in chloroplasts in eukaryotes and in thylakoid membranes in cyanobacteria and in the plasma membrane in other prokaryotes.

II. **EARLY STUDIES OF PHOTOSYNTHESIS, pp. 355-358**

    A.  **The Basic Equation of Photosynthesis was Derived from the Independent Observations of Several Scientists, pp. 355-356**
        1.  In the 1600s, Van Helmont determined that plants use inorganic nutrients.

# Chapter 9

    2. In 1771, Priestly discovered that a plant could keep a mouse alive in an anoxic environment.
    3. Ingenhousz discovered that only the green portion of the plant kept the mice alive and only in the presence of light.
    4. Senebrier discovered plants use $CO_2$ and produce $O_2$.
    5. In 1845, Mayer developed the overall equation for photosynthesis:
$$CO_2 + 2\,H_2O \xrightarrow[\text{green plant}]{\text{light}} (CH_2O)_n + O_2$$
    6. When n = 6, the carbohydrate is glucose.

  **B. Chloroplasts Carry Out Photosynthesis in Two Stages Called the Light and Dark Reactions, pp. 356-357**
    1. The light reaction uses light energy and produces oxygen; it is temperature-independent and occurs in the chloroplast membranes.
    2. $CO_2$ fixation occurs during the dark reaction; it is temperature-sensitive.
    3. Isolated chloroplasts produce oxygen in the presence of an artificial electron acceptor (the Hill reaction):
$$4\,Fe^{3+} + 2\,H_2O \xrightarrow{\text{light}} 4\,Fe^{2+} + 4\,H^+ + O_2$$

  **C. The Light and Dark Reactions of Photosynthesis Are Linked by NADPH and ATP, pp. 357-358**
    1. The light reaction produces NADPH and ATP that are used in the dark reaction.

**III. ANATOMY OF THE CHLOROPLAST, pp. 358-366**

  **A. Early Light Microscopic Studies Identified the Chloroplasts as the Site of Photosynthesis, p. 358**
    1. In 1894, Engelmann discovered chloroplasts are the site of the light reaction.

  **B. Chloroplasts Contain Membranes That Define Three Separate Compartments, pp. 358-362**
    1. The chloroplast is surrounded by the chloroplast envelope enclosing the stroma.
    2. The chloroplast envelope consists of the inner and outer membranes surrounding an intermembrane space.
    3. The thylakoid membranes are in the stroma; the space between thylakoid membranes is the thylakoid lumen.
    4. Thylakoid membranes can be stacked as grana thylakoids or unstacked as stroma thylakoids.
    5. Thylakoid membranes have $CF_1$ particles protruding from the surface.

  **C. The Main Components of the Chloroplast Can Be Isolated for Biochemical Study, p. 362**
    1. Class II chloroplasts (isolated by harsh homogenization techniques) cannot fix $CO_2$.
    2. Class I chloroplasts (isolated from protoplasts) can fix $CO_2$.

  **D. Chloroplasts Are Enclosed by a Relatively Permeable Outer Membrane and an Impermeable Inner Membrane, pp. 362-364**
    1. Porins in the outer membrane allow small molecules and ions into the intermembrane space.
    2. The inner membrane is coiled into the peripheral reticulum and contains transport proteins and enzymes.

3. The inner membrane is the site of lipid synthesis for the plant cell.

**E. The Chloroplast Stroma Is the Site of $CO_2$ Fixation, pp. 364-365**
    1. The stroma contains starch grains, plastoglobuli (lipid deposits). Algal chloroplasts contain pyrenoids (rubisco); chloroplasts of higher plants have stroma centers (which contain rubisco).

**F. Thylakoid Membranes Contain Unusual Lipids and Numerous Proteins Involved in the Light Reactions, pp. 365-366**
    1. Unlike other membranes, the thylakoid membranes consist primarily of glycolipids and chlorophyll and carotenoids.
    2. Thylakoid membranes have a high protein-to-lipid ratio.

## IV. THE LIGHT REACTIONS OF PHOTOSYNTHESIS, pp. 366-385

**A. Chlorophyll Is the Principal Light-Absorbing Pigment in Thylakoid Membranes, pp. 366-367**
    1. Photosynthetic pigments can be extracted with organic solvents and separated by chromatography.
    2. Chlorophyll is the green pigment; it is composed of a porphyrin ring containing a magnesium atom and is imbedded in the membrane by the hydrocarbon phytol.
    3. The double bonds in the porphyrin ring absorb light.
    4. Chlorophyll absorbs blue and red light.

|  | Eukaryotes | | | | Prokaryotes | | |
| --- | --- | --- | --- | --- | --- | --- | --- |
|  | Plants | Green algae | Diatoms, Brown algae | Red algae | Cyanobacteria | Purple bacteria | Green bacteria |
| Chlorophyll *a* | √ | √ | √ | √ | √ | | |
| Chlorophyll *b* | √ | √ | | | | | |
| Chlorophyll *c* | | | √ | | | | |
| Bacteriochlorophyll *a* | | | | | | √ | √ |
| Bacteriochlorophyll *b* | | | | | | √ | |
| *Chlorobium* chlorophylls | | | | | | | √ |

**B. Carotenoids and Other Accessory Pigments Funnel Light Energy to Chlorophyll, pp. 367-368**
    1. Carotenoids are lipids that absorb violet/blue-green light.
    2. Phycobilins are light-absorbing pigments consisting of four porphyrin rings in a linear chain.
    3. Phycoerythrins are bound to proteins to form phycobilisome granules.
    4. Light energy absorbed by accessory pigments is transferred to chlorophyll *a*.

|  | Eukaryotes | | | | Prokaryotes | | |
| --- | --- | --- | --- | --- | --- | --- | --- |
|  | Plants | Green algae | Diatoms, Brown algae | Red algae | Cyanobacteria | Purple bacteria | Green bacteria |
| Carotenoids | √ | √ | √ | √ | √ | √ | √ |
| Phycoerythrin | | | | √ | | | |
| Phycocyanin | | | | | √ | | |

C. **Photosystems Are Formed from a Mixture of Chloroplasts, Carotenoids, and Proteins, pp. 368-370**
   1. 2,500 chlorophyll molecules are needed to fix 1 $CO_2$.
   2. Eight photons of light are needed to fix 1 $CO_2$.
   3. Photosynthetic pigments are organized into photosystems; each photosystem has a light-harvesting complex and a reaction-center complex.
   4. When a photon is absorbed by one chlorophyll or carotenoid in a light-harvesting complex, an electron is activated.
   5. The excited state is transferred from one pigment to another until it reaches a pair of chlorophyll *a* molecules in the reaction-center complex.
   6. An excited electron is transferred to the photosynthetic electron transfer chain.

D. **The Photosynthetic Electron Transfer Chain Is Analogous to the Mitochondrial Respiratory Chain, pp. 370-371**
   1. The photosynthetic electron transfer chain carries electrons from water to $NADP^+$; it is an energy-consuming process.
   2. The electron carriers are cytochromes $b_6$ and *f*, ferredoxin, $NADP^+$ reductase, plastocyanin, plastoquinone, and $NADP^+$.

E. **Two Photosystems Are Involved in the Photosynthetic Electron Transfer Chain of Chloroplasts, p. 371**
   1. The reaction-center chlorophyll of photosystem I is P700 (700 nm).
   2. The reaction-center chlorophyll of photosystem II is P680.

F. **The Photosynthetic Electron Transfer Chain Is Organized into Several Protein Complexes That Transfer Electrons from Water to $NADP^+$, pp. 371-374**
   1. The electron carriers are arranged in a Z scheme.
   2. **Photosystem II splits water and transfers its electrons to plastoquinone, p. 372**
      a. The light-harvesting complex of photosystem II (LHCII) contains several dozen chlorophyll *a* and *b* molecules. Light is absorbed to create an excited P680*; the excited electron is transferred to pheophytin, $Q_A$, and $Q_B$.
      b. An electron is returned in P680 from the hydrolysis of water.
         $$2 H_2O \rightarrow O_2 + 4 H^+ + 4 e^-$$
      c. P680 gives up 4 electrons before any are replaced (water-oxidizing clock).
   3. Reduced $Q_B$ diffuses through the lipid bilayer to deliver its electrons to the cytochrome $b_6$-*f* complex; which passes the electrons to plastocyanin in the thylakoid lumen; plastocyanin passes the electrons to photosystem I.
   4. The light-harvesting complex of photosystem I (LHCI) contains ~100 chlorophyll *a* and *b* molecules. Light is absorbed to create an excited P700*; the excited electron is transferred to ferredoxin.
      a. P700 receives an electron from plastocyanin.
   5. Electrons are transferred from ferredoxin to $NADP^+$ by the enzyme $NADP^+$ reductase.

G. **ATP Is Synthesized by Both Noncyclic and Cyclic Photophosphorylation, pp. 374-375**
   1. The flow of electrons generates ATP by photophosphorylation.

Chapter 9

      2.    Noncyclic photophosphorylation: As electrons flow between the carriers, protons are pumped across the thylakoid membrane. The resulting electrochemical proton gradient drives ATP formation.
      3.    In cyclic photophosphorylation, electrons flow through photosystem I, then to cytochrome $b_6$-$f$ complex, then return to photosystem I by plastocyanin. NADPH and oxygen are not produced in cyclic photophosphorylation.

**H. The Photosynthetic Electron Transfer Chain Contains Two Coupling Sites for Photophosphorylation, pp. 375-376**
    1.    A coupling site is a region in an electron transfer chain where sufficient energy is released to establish a proton gradient that can be used to make ATP.
    2.    The coupling sites are located near plastoquinone and associated with the hydrolysis of water.

**I. Membrane Reconstitution Studies Have Revealed That the $CF_1$-$CF_0$ Complex Is an ATP Synthase, p. 377**
    1.    $CF_1$ protrudes from the outer (stromal) surface of the thylakoid membrane; it is attached to $CF_0$ imbedded in the membrane.
    2.    The $CF_1$-$CF_0$ complex is referred to as ATP synthase.

**J. An Electrochemical Proton Gradient across the Thylakoid Membrane Drives ATP Formation, pp. 377-383**
    1.    Protons are pumped into the thylakoid lumen during electron transfer.
    2.    Artificially created pH gradients can drive ATP synthesis in the absence of light (the thylakoid lumen is acidic relative to the stroma).
    3.    The driving force for photophosphorylation is mostly due to $\Delta$pH rather than $\Delta\psi$.
    4.    Six protons are pumped across the thylakoid membrane per pair of electrons passing from water to $NADP^+$; 3 protons are needed to drive the synthesis of 1 ATP.
    5.    $CF_0$ serves as a transmembrane channel through which protons flow.
    6.    The components of the electron transfer chain are asymmetrically oriented across the thylakoid membrane to pump protons into the thylakoid lumen.

**K. Photosystems I and II Are Spatially Separated in Stacked and Unstacked Thylakoids, pp. 383-384**
    1.    Photosystem II is located in stacked thylakoid membranes.
    2.    Photosystem I is located in unstacked thylakoid membranes.
    3.    The cytochrome $b_6$-$f$ complex is located in stacked and unstacked membranes.

**L. LHCII Acts as a Regulator of Photosystem Activity, p. 385**
    1.    When photosystem II is working faster than photosystem I, LHCII kinase activates LHCII which moves to unstacked thylakoids to allow photosystem I to absorb more light.

**V. THE $CO_2$-FIXING DARK REACTIONS, pp. 385-391**
    1.    The dark reaction of photosynthesis uses NADPH and ATP to fix $CO_2$.
    2.    The dark reaction does not *use* light, hence, the name *dark* reaction.

A. **Experiments Using Radioactive $CO_2$ First Revealed the Role Played by the Calvin Cycle in $CO_2$ Fixation, pp. 386-388**
   1. Calvin exposed green algae to $^{14}CO_2$ to identify the steps in the fixation of $CO_2$ (the Calvin Cycle).
   2. Ribulose 1,5-bisphosphate (RuBP) reacts with $CO_2$ to form 3-phosphoglycerate (3-PGA).

B. **Glyceraldehyde 3-Phosphate Is the Principal Product of $CO_2$ Fixation by the Calvin Cycle, p. 388**
   1. ATP and NADPH from the light reaction are used to convert 3-PGA to glyceraldehyde 3-phosphate.
   2. Five glyceraldehyde 3-phosphates are used to regenerate RuBP and can enter glycolysis to be converted into glucose phosphate, a precursor for cellulose and sucrose.
   3. Glyceraldehyde 3-phosphate that remains in chloroplasts to converted into starch.

C. **The Calvin Cycle Requires 18 Molecules of ATP and 12 Molecules of NADPH When Converting $CO_2$ to Glucose, pp. 388-389**
   1. $6 CO_2 + 18 ATP + 12 NADPH + 12 H^+ \rightarrow$
      $C_6H_{12}O_6 + 18 ADP + 18 P_i + 12 NADP^+ + 6 H_2O$

D. **Combining the Light Reactions with the Calvin Cycle Generates the Overall Equation of Photosynthesis, p. 389**
   1. The $\Delta G°'$ for this reaction is 686 kcal/mol.
   2. If 8 photons are needed to fix 1 $CO_2$; 48 photons are needed to make one glucose.
   3. If the $\Delta G°' = 41$ kcal/mol per photon, 1,968 kcal of light energy are used to synthesize one glucose.

E. **Molecules Involved in Chloroplast Metabolism Are Selectively Transported across the Inner Chloroplast Membrane, pp. 389-390**
   1. Triose phosphate translocator is an antiport for glyceraldehyde 3-phosphate out of the chloroplast and inorganic phosphate or 3-PGA in.

F. **Light Regulates the Activity of Rubisco and Other Calvin Cycle Elements, pp. 390-391**
   1. Rubisco catalyzes the reaction: $CO_2$ + RuBP.
   2. Rubisco is activated by $Mg^{2+}$ and pH (8).
   3. Rubisco is carboxylated by rubisco activase in the presence of light.

VI. **PHOTORESPIRATION AND $C_4$ PLANTS, pp. 391-396**
   1. Photorespiration is a light-dependent pathway that uses $O_2$ and produces $CO_2$.
   2. $CO_2$ fixation is inhibited by increasing concentration of $O_2$.

A. **Photorespiration Occurs When $O_2$ Substitutes for $CO_2$ in the Reaction Catalyzed by Rubisco, pp. 391-392**
   1. When $[O_2]$ is high and $[CO_2]$ is low, rubisco catalyzes:
      RuBP + $O_2$ → phosphoglycolate and 3-PGA
   2. Phosphoglycolate is transferred to peroxisomes to combine with $O_2$ to produce glycine.

3. The glycine goes to the mitochondria where
   2 glycine → serine + $CO_2$
4. The serine is returned to the peroxisome for conversion into glycerate which returns to the Calvin cycle in the chloroplast.
5. Photorespiration generates $CO_2$ from RuBP without any of the energy released in a useful form.

**B. $C_4$ Plants Minimize Photorespiration by Using a Special Pathway for Concentrating $CO_2$, pp. 392-394**
1. To minimize photorespiration, some plants keep a high [$CO_2$] concentration in photosynthesizing cells.
2. Plants that depend solely on the Calvin cycle are called $C_3$ plants because they produce 3-PGA (a *3* carbon compound).
3. $C_4$ plants produce oxaloacetate and malate (*4* carbon compounds) from $CO_2$ fixation.
4. Bundle sheath cells surround the veins in the leaves of $C_4$ plants.
5. In the mesophyll cells, PEP carboxylase joins $CO_2$ to PEP to form oxaloacetate (and later malate).
6. The four-carbon acids are transported to bundle sheath cells and degraded to $CO_2$ and pyruvate.
7. In the bundle sheath cells, the $CO_2$ can be fixed in the Calvin cycle.
8. $C_4$ plants live in hot, dry environments and must keep their stomata closed during the day. This limits entry of $CO_2$ into the leaves.
9. The stomata are closed during the day to prevent water loss and so $O_2$ can't compete with rubisco.

**C. CAM Plants Produce Malate at Night and Use It to Provide $CO_2$ to the Calvin Cycle, pp. 394-396**
1. CAM (crassulacean acid metabolism) live in hot, dry environments.
2. CAM plants open their stomata at night and $CO_2$ is fixed by PEP carboxylase in malate acid in mesophyll cells.
3. During the day, malate is broken down in the cytosol to release $CO_2$.
4. The $CO_2$ can be fixed by rubisco.
5. The stomata are closed during the day to prevent water loss and so $O_2$ can't compete with rubisco.

**VII. PHOTOSYNTHESIS IN PROKARYOTES, pp. 396-398**
1. In cyanobacteria, the light reactions occur in thylakoid membranes distributed throughout the cytoplasm; most of the light energy is absorbed by phycobilin and passed to chlorophyll.
2. Cyanobacterial photosynthesis is like plant photosynthesis (producing $O_2$ and NADPH).

**A. The Light Reactions in Purple and Green Bacteria Utilize Only One Photosystem, p. 396**
1. The photosynthetic membrane of purple bacteria is the plasma membrane folded into vesicles.
2. The photosynthetic membranes of green bacteria are spherical vesicles, not associated with the plasma membrane.
3. The reaction center is P870.
4. When bacteriochlorophyll *b* absorbs a photon of light, it transfers an electron to bacteriophaeophytin; which passes an electron to $Q_A$, and on to $Q_B$.

5. When $Q_B$ has received 2 electrons (and 2 protons from solution), it diffuses to cytochrome $b$-$c_1$ complex. From the cytochrome complex, the electrons are transferred back to P870.
6. At the cytochrome complex, protons are pumped into the lumen of the photosynthetic membranes causing an electrochemical gradient that can drive the formation of ATP.

**B. Photosynthetic Bacteria Can Make NADPH Using Electron Donors Other Than Water, p. 398**
1. Electrons can be transferred through carriers to $NADP^+$ instead of being returned to P870.
2. Electrons can be returned to P870 by $H_2S$.
3. The overall equation for anoxygenic prokaryotic photosynthesis is:
   $6 CO_2 + 12 H_2S \rightarrow C_6H_{12}O_6 + 6 H_2O + 12 S$

**C. *Halobacterium* Captures Solar Energy without Fixing $CO_2$ or Other Electron Acceptors, p. 398**
1. *Halobacterium* has the purple pigment bacteriorhodopsin in its plasma membrane.
2. Bacteriorhodpsin absorbs light at 560 nm.
3. Light causes bacteriorhodopsin to pump protons creating a proton gradient that drives ATP formations.

# KEY TERMS

| | | |
|---|---|---|
| ATP synthase | autotroph | bacteriorhodopsin |
| bundle sheath cell | $C_3$ plant | $C_4$ plant |
| Calvin cycle | CAM | carotenoid |
| $CF_1$ particle | $CF_1$-$CF_0$ complex | chlorophyll |
| chloroplast | chloroplast envelope | $CO_2$ fixation |
| coupling site | cyclic photophosphorylation | dark reaction |
| etioplast | grana | grana thylakoid |
| heterotroph | LCHI | LHCII |
| light reaction | light-harvesting complex | mesophyll cell |
| noncyclic photophosphorylation | P680 | P700 |
| PEP carboxylase | pheophytin | photon |
| photophosphorylation | photorespiration | photosynthesis |
| photosystem | photosystem I | photosystem II |
| phycobilisome granules | protoplast | pyrenoid |
| $Q_A$ | $Q_B$ | reaction-center complex |
| rubisco | starch | stomata |
| stroma | stroma center | stroma thylakoid |
| thylakoid lumen | thylakoid membrane | water-oxidizing clock |

# Chapter 9

## 👉 KEY FIGURE

Label the components of photosystem I and photosystem II. Diagram the path of electrons and protons. Label ATP synthase and the acidic side of the membrane.

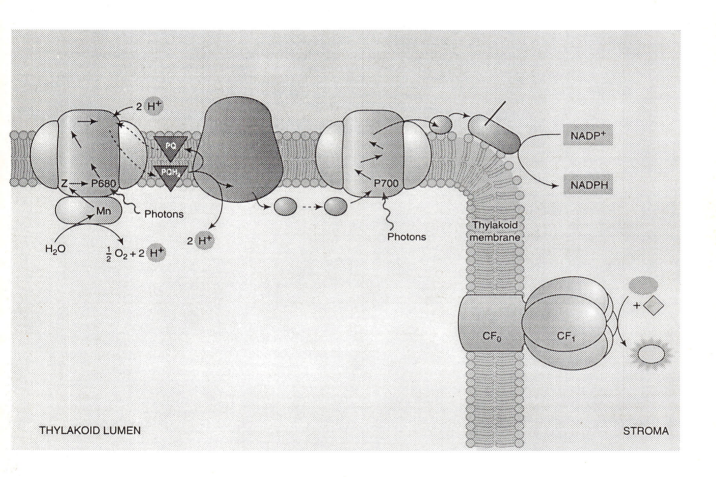

# Chapter 9

## 👉 STUDY QUESTIONS

1. $CF_1$ particles isolated from the thylakoid membrane can hydrolyze ATP. This suggests that $CF_1$ particles
   a. Transfer electrons from water to $NADP^+$
   b. Synthesize ATP
   c. Synthesize ATP in the presence of light
   d. Are responsible for respiratory electron transfer
   e. None of the above

Use the following diagram to answer questions 2-5.

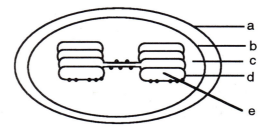

2. Location of photosystem II.

3. Location of porins.

4. Location of plastocyanin.

5. Site of $CO_2$ fixation.

6. After leaving P680, where does an electron go next?
   a. Ferredoxin
   b. $NADP^+$
   c. P700
   d. Plastocyanin
   e. Plastoquinone

7. Your lab partner observes a structure which has porins and thylakoid membranes. Which of the following are you sure that it also has?
   a. Bacteriochlorophyll *a*
   b. Chlorophyll *a*
   c. Chlorophyll *b*
   d. Phycoerythrin
   e. None of the above

Use the following choices to answer questions 8-13:
   a. Cyclic photophosphorylation
   b. Noncyclic photophosphorylation
   c. Both light reactions
   d. Calvin cycle
   e. None of the above

8. Which process produces $CO_2$?

9. Which process is a series of redox reactions?

10. Which process uses light-energized electrons?

11. Which process produces NADPH?

12. Which process produces phosphoglyceraldehyde?

13. In which process is chlorophyll the initial electron donor and the ultimate electron acceptor?

14. All of the following occur in photosynthesis and respiration **except**
    a. Electron flow
    b. Splitting of water molecules
    c. Synthesis of ATP
    d. Establishing a proton gradient
    e. Electron carriers embedded in a membrane

15. Electrons are replaced in P700 from
    a. Water.
    b. NADPH.
    c. Plastocyanin.
    d. Chlorophyll *b*.
    e. Photosystem I.

16. Which of the following is not required at some stage during the process of photosynthesis?
    a. ADP
    b. $O_2$
    c. Water
    d. ATP
    e. $NADP^+$

17. Which of the following processes does not generate ATP?
    a. Photophosphorylation
    b. Calvin Cycle
    c. Oxidative phosphorylation
    d. Substrate-level phosphorylation
    e. None of the above

18. All of the following conditions are necessary for rubisco to combine with $CO_2$ **except**
    a. $Mg^{2+}$.
    b. $O_2$.
    c. pH 8.
    d. Light.
    e. Rubisco activase.

19. Electrons pass through the following sites in the Z-scheme for photosynthesis. Which is the first step?
    a. Cytochrome $b_6$-$f$ complex
    b. NADPH
    c. P680
    d. P700
    e. Plastoquinone

20. If you supplied a plant with $^3H_2O$, where would you find the $^3H$ in the products of photosynthesis?
    a. Oxygen
    b. Carbon dioxide
    c. Water
    d. Glucose
    e. Chlorophyll

21. All of the following are true about photorespiration **except** that
    a. It generates ATP.
    b. It occurs in $C_3$ plants.
    c. It occurs in temperate-zone plants.
    d. CAM plants avoid it by closing their stomata during the day
    e. It involves rubisco

## PROBLEMS

1. Engelmann illuminated different chloroplasts of the green alga *Spirogyra* with colored light, then observed that aerobic bacteria congregated around one of the chloroplasts. Explain what happened.

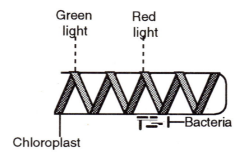

2. Explain why only red algae and not green or brown algae are found below about 75 m. in the ocean.

3. Why can't most organisms live in 15% NaCl? How does the purple membrane of *Halobacterium* bacteria enable these bacteria to live in a 15%-NaCl environment?

4. DCMU (dichlorophenyl dimethyl urea) uncouples photosystems I and II. What effect would an herbicide containing DCMU have if sprayed on a broccoli field to eliminate weeds?

5. *Rhodopseudomonas* bacteria are anaerobic autotrophs which use organic compounds as an electron donor. Diagram the metabolic pathways of this bacterium.

# Chapter 9

6. The absorption spectrum of *Chlamydomonas* is shown below. Explain the burst of oxygen in the red region.

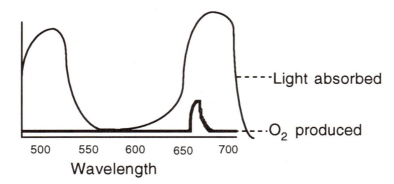

7. Assume that you carried out a series of experiments using a suspension of intact chloroplasts and an artificial electron acceptor, DCIP (dichlorophenol-indophenol). DCIP is blue when it is oxidized and turns colorless when it is reduced. What conclusions can you draw from these data?

| Light intensity (ft-candles) | Incubated 4 min. in white light, 20°C | DCIP color | The other team's DCIP color* |
|---|---|---|---|
| 1000 | | Colorless | Blue |
| 500 | | Colorless | Blue |
| 250 | | Bluish tinge | Blue |
| 125 | | Blue | Blue |
| Color of light | Incubated 4 min. at 1000 ft-candles of light, 20°C | DCIP color | The other team's DCIP color* |
| White | | Colorless | Blue |
| Red | | Colorless | Blue |
| Green | | Blue | Blue |
| Blue | | Colorless | Blue |
| Temperature (°C) | Incubated 4 min. at 1000 ft-candles of light | DCIP color | The other team's DCIP color* |
| 15 | | Colorless | Blue |
| 20 | | Colorless | Blue |
| 37 | | Colorless | Blue |
| 45 | | Colorless | Blue |

*Another lab team performed these same experiments but prepared their chloroplasts in a hypotonic solution with acetone. Explain their results.

## ANSWERS TO STUDY QUESTIONS

1. b    2. d    3. a    4. e

5. c    6. e    7. b    8. e

9. c    10. c    11. b    12. d

13. b    14. b    15. c    16. b

17. b    18. b    19. c    20. d

21. a

# CHAPTER 10

## *The Nucleus and Transcription of Genetic Information*

### ⇗ LEARNING OBJECTIVES

Be able to
1. Describe the structure of the nucleus, including the nuclear lamina and nuclear pore complexes.
2. Describe the structure of eukaryotic chromatin, including the structure of the nucleosome.
3. Compare the packaging of eukaryotic and prokaryotic DNA into a cell.
4. Differentiate between RNA polymerase I, RNA polymerase II, and RNA polymerase III.
5. Define TATA and CCAAT boxes.
6. Outline the steps in gene transcription, include the preinitiation complex and termination events.
7. Outline the events in RNA processing in eukaryotes.
8. Describe how transcription of an inducible operon and a repressible operon are controlled.
9. Briefly explain the C value paradox using the following terms:
    a. Interspersed repeats
    b. Satellite DNAs
    c. Interbands
    d. Heterochromatin
10. List five characteristics of transcriptionally active DNA in eukaryotic cells.
11. List five ways transcription can be activated in a eukaryotic cell.
12. Compare and contrast DNA binding by each of the following:
    a. Helix-turn-helix motif
    b. Zinc fingers
    c. Leucine zipper
    d. Helix-loop-helix motif

### ⇗ CHAPTER OVERVIEW

I. THE NUCLEUS AND THE NUCLEOID, pp. 402-417

  A. The Nucleus of Eukaryotic Cells is Bounded by a Double-Membrane Envelope, pp. 402-403
    1. The double-membrane nuclear envelope surrounds the nucleoplasm.
    2. The nucleoplasm contains the nucleoli and chromatin.
    3. The nuclear matrix supports the nucleus.

  B. The Nuclear Envelope Is Supported by the Nuclear Lamina and Contains Numerous Nuclear Pore Complexes, pp. 403-404
    1. The perinuclear space is between the inner and outer nuclear membranes.
    2. The outer nuclear membrane is continuous with the ER and may have ribosomes on the outer surface.
    3. The perinuclear space is continuous with the lumen of the ER.
    4. The nucleus is held in place by the cytoskeleton.
    5. The nuclear lamina forms a thin layer beneath the inner nuclear membrane.

# Chapter 10

      6.    Nuclear pore complexes (NPC) form openings in the nuclear envelope.
      7.    An eight-sided NPC consists of a central transporter attached to nucleoplasmic and cytoplasmic rings.

C. **Nuclear Pore Complexes Are Channels through Which Molecules Enter and Exit the Nucleus, pp. 404-407**
    1.    Small particles (< 10 nm) pass through nuclear pores by diffusion.
    2.    The central transporter channel acts as a gated channel to transport materials up to 25 nm in diameter.
    3.    Molecules are targeted to enter the nucleus by a short amino acid sequence called the nuclear localization sequence.

D. **The Nucleoplasm Is the Soluble Portion of the Nucleus, p. 407**
    1.    The nucleoplasm contains ions, coenzymes, and nucleotides; proteins (enzymes); and RNA molecules, mostly in ribonucleoprotein complexes.

E. **The Nucleolus Contains Granules and Fibrils Involved in Ribosome Formation, p. 407**

F. **The Nuclear Matrix Provides a Supporting Framework for the Nucleus, p. 408**
    1.    The nuclear matrix may be important in anchoring chromatin fibers where DNA or RNA is being synthesized.

G. **Chromatin Fibers Consist of DNA Associated with Histone and Nonhistone Proteins, pp. 408-409**
    1.    DNA and its associated proteins dispersed throughout the nucleus is called chromatin; when condensed it is called chromosomes.

H. **Histones Are Small Basic Proteins That Are Similar in All Eukaryotic Cells, pp. 409-410**
    1.    Histones have a high lysine and arginine content giving them a + change.
    2.    There are five types of histones: H1, H2A, H2B, H3, and H4.
    3.    Histones are present in all eukaryotes.
    4.    The sperm of some animals contain protamines instead of histones.

I. **Nonhistone Proteins Are a Heterogeneous Group of DNA-Associated Molecules, pp. 410-411**
    1.    Nonhistone proteins include DNA and RNA polymerases and high-mobility group (HMG) proteins.
    2.    HMG proteins are associated with transcription.

J. **The Nucleosome Is the Fundamental Unit of Chromatin Structure, pp. 411-412**
    1.    DNA may be 1000 times longer than the nucleus and therefore must remain tightly packed in the nucleus
    2.    Histones appear on the chromosome at 10 nm intervals, or 200 base pairs.
    3.    The 200 base pairs of DNA associated with a protein particle is called a nucleosome.

K. **An Octamer of Histones Forms the Core of the Nucleosome and Is Associated with 146 Base Pairs of DNA, pp. 412-414**
    1.    Two H2A-H2B dimers and two H3-H4 dimers form a histone.

2. The core particle of the nucleosome consists of the octamer and 146 base pairs.
3. The DNA wraps around the core particle ~1.8 times.
4. The linker DNA (~54 base pairs) joins the nucleosomes.

**L. Nucleosomes Are Packed into Higher-Order Structures to Form Chromatin Fibers and Chromosomes, pp. 414-415**
1. The packing ratio = length of the DNA ÷ length of the particle into which the DNA is packaged.
2. The chain of nucleosomes coils into a solenoid mediated by H1.

**M. The Bacterial Nucleoid Contains DNA That Is Extensively Supercoiled and Bound to a Few Basic Proteins, pp. 415-417**
1. DNA in bacterial cells is packaged into a region called a nucleoid.
2. The bacterial chromosome is a single, circular DNA molecule.
3. The DNA is supercoiled to fit into the cell; when DNA is twisted the same way as the helix it is positive supercoiling. Negative supercoiling is when the DNA is twisted in the opposite direction.
4. Topoisomerses control supercoiling.
5. Bacterial DNA is associated with some proteins.

## II. GENE TRANSCRIPTION, pp. 417-425

**A. RNA Is Synthesized by Several Different Kinds of RNA Polymerase, pp. 417-418**
1. In prokaryotic cells, a single type of RNA polymerase makes mRNA, rRNA, and tRNA.
2. The RNA polymerase consists of a core enzyme and the sigma factor.
3. Eukaryotic cells contains 3 RNA polymerases.
4. RNA polymerase I makes rRNA.
5. RNA polymerase II makes mRNA.
6. RNA polymerase III makes tRNA and 5S rRNA.

**B. RNA Polymerase Binds to DNA Promoter Sites, pp. 418-422**
1. Promoters are written in the 5′→ 3′ direction on the strand complementary to the transcription template.
2. In bacteria, the consensus sequences in promoters are TATAAT located 10 bases upstream from the start site and TTGACA located 35 bases upstream.
3. In eukaryotes, the promoters for RNA polymerase II contain the consensus sequences called TATA box (TATAAA, 30 bases upstream), CCAAT box, and GC box (GGGCG).
4. The promoter for RNA polymerase III is located in the middle of the gene; this is an internal control region.

**C. Transcription Factors Allow RNA Polymerase to Recognize Promoter Sites and Initiate RNA Synthesis, pp. 422-423**
1. In prokaryotes, different sigma factors recognize different promoter sequences.
2. In eukaryotes, transcription factors bind to the promoter and/or RNA polymerase to form a preinitiation complex.
3. Response elements are DNA sequences located upstream from the promoter. Transcription factors binding to response elements regulate transcription.

# Chapter 10

    **D. Enhancers and Silencers Alter the Activity of Promoters That May Be Located Far Away in the DNA Molecule, p. 423**
        1. Enhancers are DNA sequences that stimulate transcription.
        2. Silencers inhibit transcription.
        3. The DNA is folded so proteins bound to enhancers or silencers can interact with the preinitiation complex.

    **E. RNA Synthesis is Usually Initiated with ATP or GTP after the DNA Template Has Been Partially Unwound, p. 423**
        1. RNA polymerase unwinds about 17 base pairs of DNA.
        2. The first base added is ATP or GTP.

    **F. Elongation of the RNA Chain Occurs in the 5′→ 3′ Direction, pp. 423-424**
        1. Nucleotides are added by forming a phosphodiester bond between the first phosphate group and the 3′ hydroxyl of the preceding base.
        2. Two phosphates are released as pyrophosphate.

    **G. Transcription Can Be Terminated by RNA Sequences That Either Form a Stem-and-Loop Structure or Are Recognized by Rho Factor, pp. 424-425**
        1. In bacteria, termination can occur when a G-C sequence is transcribed and folds on itself.
        2. The rho factor separates mRNA from DNA.
        3. An antitermination protein can suppress termination of rho-dependent genes.
        4. In eukaryotes, transcription is not formally terminated and can continue far beyond the gene.

**III. PROCESSING OF MESSENGER RNA, pp. 425-432**
        1. Prokaryotic RNA transcripts may contain more than one gene (polycistronic).
        2. Eukaryotic RNA transcripts may need to be cleaved into mature RNA (RNA processing).
        3. In prokaryotes, only rRNA and tRNA are processed.

    **A. Most of the RNA Sequences Synthesized in Eukaryotic Cells Are Degraded in the Nucleus without Entering the Cytoplasm, pp. 426-427**
        1. Less than 10% of the RNA made in the nucleus gets to the cytoplasm; 90% is degraded in the nucleus.
        2. Nuclear RNA is called heterogeneous nuclear RNA (hnRNA).

    **B. Messenger RNAs Are Derived from Larger RNA Precursors, p. 427**
        1. hnRNAs serve as pre-RNAs.
        2. Pre-RNAs are selectively cleaved and spliced to produce mRNA.

    **C. Messenger RNA Precursors Are Packaged with Protein Particles as They Are Being Transcribed, pp. 427-428**
        1. Pre-RNAs are bound to proteins to form nuclear ribonucleoprotein particles.

    **D. Messenger RNA Precursors Acquire 5′ Caps and 3′ Poly-A Tails, pp. 428-429**
        1. Ribose of the first and second bases in pre-RNA are methylated to form the 5′cap.
        2. After transcription, the pre-RNA is cleaved downstream from an AAUAAA sequence and 50-250 adenine nucleotides are added to the 3′ end of the pre-RNA.

- E. **Eukaryotic Genes Are Interrupted by Introns That Must Be Removed During RNA Processing, pp. 429-430**
    1. DNA sequences coding for mRNA are called exons; exons are separated by introns.

- F. **RNA-Protein Complexes Called snRNPs Facilitate the Removal of Introns from Pre-mRNA, pp. 430-432**
    1. RNA splicing removes introns from pre-RNA.
    2. The 5' end of an intron starts with GU; the 3' end ends with AG.
    3. A group of small nuclear ribonucleoprotein particles bind to an intron to form a spliceosome.
    4. The 5' end of the slice site is cleaved and the 5' end of the intron is joined to the branch point to form a lariat. (It looks like a loop.)
    5. The 3' end of the splice site is cleaved and the two ends of the exon are joined together.
    6. The released intron is degraded.
    7. A few genes have been found to be self-slicing; the intron removal occurs without snRNPs.

IV. **GENE REGULATION IN PROKARYOTES, pp. 432-441**

- A. **Genes Involved in Lactose Metabolism Are Organized into an Inducible Operon, pp. 432-433**
    1. The *lac* operon is an inducible operon.
    2. Induction occurs when a small molecule (called an inducer) triggers the production of a protein.
    3. The *lac* operon consists of structural genes (*lac*Z, *lac*Y, and *lac*A), a promoter (*lac*P), and an operator (*Lac*O).
    4. *Lac*I codes for a repressor.

- B. **The *Lac* Repressor Is a Protein That Contains Binding Sites for Both DNA and Inducers Such as Lactose, pp. 434-435**
    1. The repressor binds to the operator inhibiting transcription.
    2. When lactose binds to the repressor, the operator is free and transcription can occur.

- C. **The *Lac* Repressor Inhibits Transcription by Binding to *Lac* Operator DNA through a Helix-Turn-Helix Motif, pp. 435-436**
    1. The amino acids in one α helix form hydrogen bonds with DNA bases.
    2. The *lac* operator is ~24 bases in a palindrome.

- D. **The Catabolite Activator Protein (CAP) and Cyclic AMP Exert Positive Control Over Operons Susceptible to Catabolite Repression, pp. 436-437**
    1. The *lac* repressor exerts negative control on the operon.
    2. A DNA-binding protein that activates transcription exerts positive control.
    3. The presence of glucose inhibits the synthesis of catabolic enzymes to degrade other substrates by catabolite repression.
    4. Glucose causes a decrease in [cAMP].
    5. When glucose is not available, [cAMP] increases and the catabolic activator protein binds to cAMP and then binds to DNA upstream from the promoter.
    6. RNA polymerase then binds to start transcription.

## Chapter 10

- E. **Genes Coding for Enzymes Involved in Tryptophan Synthesis Are Organized into a Repressible Operon, pp. 437-439**
  1. Repressible operons are inhibited when a particular compound is added.
  2. The *trp* operon consists of five structural genes and a regulatory gene, *trp*R.
  3. The repressor must bind to tryptophan (the co-repressor) in order to bind to DNA.

- F. **DNA Supercoiling and DNA Sequence Rearrangements Can Both Influence Bacterial Gene Expression, pp. 439-441**
  1. Topoisomerase II catalyzes the formation of negative supercoils so transcription can occur.
  2. Inhibition of topoisomerase II will decrease transcription.
  3. In *Salmonella* bacteria, a control segment can be excised and reinserted in the opposite direction. In one position, one gene for flagellin is transcribed; in the other position, a different flagellin gene is transcribed.

- G. **Bacterial Gene Expression Can Be Regulated at Many Other Levels, p. 441**
  1. Modification of RNA polymerase will cause it to selectively transcribe genes.
  2. Transcription is also controlled by attenuation.

V. **GENE REGULATION IN EUKARYOTES, pp. 441-446**

- A. **Eukaryotic Cells Contain Far More DNA Than Appears To Be Needed for Genetic Coding Purposes, pp. 441-443**
  1. The total DNA in a haploid set of chromosomes is the C value.
  2. The C value paradox is the incongruity between the C value and the necessary amount of DNA.

- B. **DNA Reassociation Studies Reveal That Eukaryotic DNA Contains Both Unique and Repeated Sequences, pp. 443-444**
  1. About 40% of the DNA consists of repeated DNA sequences that are present in multiple copies.
  2. The remaining 60% are single copies or unique-sequence DNA.

- C. **Eukaryotic Cells Contain Several Classes of Repeated DNA Sequences, pp. 444-445**
  1. Hundreds of copies of genes for histones and rRNAs occur in cells.
  2. Satellite DNAs are short sequences that are repeated over and over accounting for 10% of the organism's DNA; they are not transcribed; they are associated with heterochromatin, centromeres, and telomeres; they may help maintain chromatin in an inactive state.
  3. Short and long interspersed repeats are not translated.

- D. **The Cytoplasm Influences the Activity of Nuclear Genes, pp. 445-446**
  1. Nuclear transplantation and cell fusion studies have shown that factors in the cytoplasm influence nuclear activity.

- E. **The Activity of Individual Genes Can Be Monitored in Polytene Chromosomes, pp. 446-447**
  1. Giant, multistranded (polytene) chromosomes occur in diptera (fly) tissues, these chromosomes form by increasing their size rather than number of cells.
  2. The cell's DNA is duplicated about 10 times to form a polytene chromosome.

3. A polytene chromosome consists of chromatin fibers aligned parallel.
4. Dark staining bands in polytene chromosomes correspond to the location of a few genes. Genes are not located in the light interbands.

F. **Chromosome Puffs and Balbiani Rings Represent Sites of Gene Transcription, pp. 447-448**
1. Some bands uncoil into a puff or into a large Balbiani ring.
2. Transcription can occur in bands that lack puffs.

G. **Chromosome Puffing Can Be Induced by the Steroid Hormone Ecdysone, pp. 448-449**
1. A puff forms as nonhistone proteins accumulate on a chromosome.
2. Early puffs code for proteins required for the development of late puffs.

H. **The Formation of Heterochromatin Allows Large Segments of DNA To Be Transcriptionally Inactivated, pp. 450-451**
1. Loosely packed, transcribable DNA is called euchromatin.
2. DNA that remains coiled is called heterochromatin and is not transcribed.
3. Constitutive heterochromatin is permanently inactive.
4. Facultative heterochromatin becomes inactive as cells specialize.
5. The Barr body in females is an X chromosome that has been converted to heterochromatin.
6. Descendants of a cell will have the same X chromosome inactivated.

I. **Transcriptionally Active Chromatin Is Sensitive To Digestion with DNase I, pp. 451-452**
1. DNA that is being transcribed is arranged in nucleosomes.

J. **DNase I Hypersensitive Sites Are Short Nucleosome-Free Regions Located Adjacent to Active Genes, pp. 452-453**
1. B DNA is a smooth right-handed double helix; Z-DNA is a left-handed helix.
2. Z-DNA has been associated with DNase I hypersensitive sites and with transcription.

K. **Changes in Histones, HMG Proteins, and Connections to the Nuclear Matrix Are Observed in Active Chromatin, pp. 453-455**
1. Acetylated histones (H3 and H4) are associated with the nucleosomes of active genes.
2. H1 is missing in active chromatin.
3. Active chromatin has a high HMG-protein content.
4. Active DNA is attached to the nuclear matrix by matrix attachment regions.

L. **DNA Methylation Suppresses the Ability of Some Eukaryotic Genes To be Transcribed, pp. 455-456**
1. Cytosine in a -CG- sequence complementary to a -GC($CH_3$)- is methylated by DNA methylase to produce 5-methylcytosine.
2. Genes with methylated cytosine are inactive.

M. **Gene Amplification and Deletion Can Alter Gene Expression, pp. 456-457**
1. Cells can produce extra copies of certain genes by gene amplification.
2. Unneeded genes can be deleted.

- **N. DNA Rearrangements Control Mating Type in Yeast and Antibody Production in Animals, p. 457**
    1. Yeast can insert different genes (for mating type) in the same location (mating type locus) by the cassette mechanism.
    2. Genes for antibodies are rearranged to produce different antibodies.
    3. Antibody-gene rearrangement brings the promoter and enhancer sequences close enough to initiate transcription.

- **O. DNA Response Elements Allow Nonadjacent Genes To Be Regulated as a Unit, pp. 457-458**
    1. DNA sequences that allow transcription to be regulated in response to a particular type of signal are called response elements.

- **P. The Heat-Shock Response Element Coordinates the Expression of Genes That Are Activated by Elevated Temperatures, p. 458**
    1. In *Drosophila*, hsp70 protein is produced when the temperature is raised.
    2. The heat-shock response element is located 62 bases upstream from the start site for transcription.
    3. A gene-specific transcription factor binds upstream from the response element.
    4. The same response element can be located near genes on different chromosomes so one stimulus can activate several genes.

- **Q. Homeotic Genes Regulate Embryonic Development by Coding for Transcription Factors Exhibiting a Helix-Turn-Helix Motif, pp. 458-459**
    1. A 180-base sequence called the homeobox codes for 60 amino acids called the homeodomain.
    2. Homeotic genes code for transcription factors that share the homeodomain; the transcription factors activate or inhibit transcription of other genes by binding to the homeobox.

- **R. Cysteine-Histidine Zinc Fingers Are Present in Transcription Factors in Variable Numbers, pp. 459-461**
    1. The promoter for RNA polymerase III is in the middle of the gene.
    2. Transcription factor TFIIIA binds to the promoter.
    3. TFIIIA consists of nine repeats of a loop (zinc finger) joined by cysteine-Zn-histidine.
    4. TFIIIA binds to DNA by these zinc fingers.

- **S. Steroid Hormones Interact with Receptors Localized in the Nucleus, p. 461**
    1. The nucleus also has receptors for thyroxin and retinoic acid.

- **T. Steroid-Hormone Receptors Act as Transcription Factors That Bind to Hormone Response Elements, pp. 461-462**
    1. Steroid hormones bind to receptors in the nucleus; the receptors act as gene-specific transcription factors.
    2. The receptor binds to DNA sequences called hormone response elements.
    3. The hormone response elements are palindromes located upstream from the gene's promoter.

# Chapter 10

- **U. The Cysteine-Cysteine Zinc Finger Is the DNA-Binding Motif of Steroid Hormone Receptors, pp. 462-463**
  1. Steroid hormone receptors bound to hsp90 and p59 cannot bind DNA.
  2. The steroid hormone displaces hsp90 and p59.
  3. Two hormone-receptor complexes bind to form a dimer.
  4. The dimer binds DNA and removes a nucleosome.
  5. Cysteine-zinc-cysteine fingers of the receptor bind the DNA.
  6. Receptors can stimulate transcription of one protein while inhibiting the transcription of others.

- **V. The Leucine Zipper and Helix-Loop-Helix Motifs Facilitate DNA-Binding by Joining Two Polypeptide Chains Together, pp. 463-464**
  1. Some DNA binding proteins contain leucines as every seventh amino acid.
  2. Two α helices containing this alternating leucine arrangement will join to form a leucine zipper.
  3. Near the leucine zipper is a region of basic amino acids that can bind to acidic DNA.
  4. Proteins Jun and Fos are joined as a leucine zipper to form the AP1 transcription factor; AP1 binds DNA to cause cell growth and division.
  5. Helix-loop-helix binding proteins contain a region of basic amino acids that can bind to acidic DNA.

- **W. Transcription Factors Contain Acidic, Glutamine-Rich, or Proline-Rich Regions That Function as Activation Domains, p. 464**
  1. The activation domain of the transcription factor is not the DNA-binding region.
  2. The activation domain makes contact with the preinitiation complex to activate transcription.

- **X. How Is the Activity of Transcription Factors Regulated? pp. 464-465**
  1. Phosphorylation of transcription factors by protein kinase A causes conformational changes that allow the transcription factor to bind to DNA.

- **Y. Control of RNA Splicing and Nuclear Export Represent Additional Ways of Regulating Eukaryotic Gene Expression, pp. 465-466**
  1. Pre-RNA can be spliced different ways to produce different mRNAs.
  2. Export of mRNA from the nucleus may require at least one intron and/or the presence of specific protein.

## ☞ KEY TERMS

5′ cap
AP1 transcription factor
C value paradox
CCAAT box
co-repressor
cysteine-cysteine zinc finger
downstream
exon
gene amplification
helix-loop-helix
heterogeneous nuclear RNA

activation domain
Balbiani ring
catabolite activator protein
chromatin
consensus sequence
DNA methylation
enhancer
Fos
gene-specific transcription
helix-turn-helix motif
histone

antitermination protein
C value
catabolite repression
chromosome
CREB protein
DNase I hypersensitive site
euchromatin
GC box
heat-shock gene
heterochromatin
hnRNP

# Chapter 10

| | | |
|---|---|---|
| homeobox | homeodomain | homeotic gene |
| inducer | internal control region | intron |
| Jun | *lac* operon | leucine zipper |
| negative control | nonhistone protein | nuclear envelope |
| nuclear lamina | nuclear localization sequence | nuclear matrix |
| nuclear pore complex | nucleoid | nucleolus |
| nucleoplasm | nucleosome | packing ratio |
| palindrome | poly-A tail | polycistronic |
| polytene | positive control | pre-RNA |
| preinitiation complex | protamine | puff |
| R looping | relaxed DNA | repeated DNA sequences |
| repressor | response element | rho |
| RNA polymerase | RNA processing | RNA splicing |
| self-splicing | sigma factor | silencer |
| snRNPs | solenoid | spliceosome |
| supercoiling | TATA box | topoisomerase |
| transcription factor | unique-sequence DNA | upstream |

## ☞ KEY FIGURE

Label the promoter, operator, RNA polymerase binding site, repressor, cAMP, and lactose. Under which conditions are the structural genes transcribed?

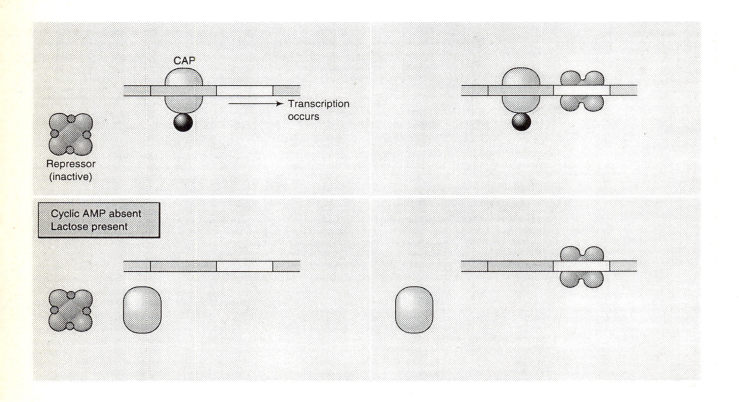

138

# Chapter 10

## STUDY QUESTIONS

1. If the nucleus of *Acetabularia mediterranea* is transplanted onto the stalk of a decapitated *A. crenulata*, which of the following cells will result?

   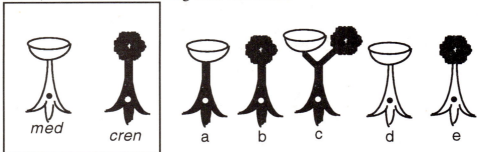

2. Maurice Wilkins observed that chromatin looks like beads on a string. This appearance is due to
   a. Ribosome synthesis.
   b. Condensation during cell division.
   c. The presence of histones.
   d. Supercoiling.
   e. None of the above.

3. Roger Kornberg discovered that nucleosomes can be prepared by mixing DNA with
   a. H2A, H2B, H3, and H4
   b. H1, H2, H3, and H4
   c. H1, H2A, H2B, and H4
   d. Micrococcal nuclease, H1, H2, and H3
   e. None of the above

4. Which one of the following is **never** transcribed?
   a. Euchromatin
   b. Constitutive heterochromatin
   c. Facultative heterochromatin
   d. Heterochromatin
   e. ~~Negatively supercoiled~~ Nucleosomal DNA

5. The following events occur during transcription. What is the third step?
   a. RNA polymerase binds to the preinitiation complex.
   b. Transcription factor binds to the promoter.
   c. RNA polymerase unwinds the DNA.
   d. mRNA grows in the 5′ → 3′ direction.
   e. The first nucleotide is added.

6. Assume that cells are incubated with $^3$H-uridine for a few minutes and then RNA synthesis is inhibited. You immediately measure the amount of $^3$H-RNA in the nucleus and cytoplasm. If you measure the amount of $^3$H-RNA in the nucleus and cytoplasm after a few hours, you would find
   a. All the $^3$H-RNA in the nucleus.
   b. All the $^3$H-RNA in the cytoplasm.
   c. 10% of the $^3$H-RNA in the cytoplasm and none in the nucleus.
   d. 10% of the $^3$H-RNA in the nucleus and 90% in the cytoplasm.
   e. 10% of the $^3$H-RNA in the cytoplasm and 90% in the nucleus.

7. Which one of the following is **not** a promoter for RNA polymerase II?
   a. TATA box
   b. CCAAT box
   c. GC box
   d. TATAAT
   e. None of the above

8. The following steps occur in RNA processing. What is the ~~fourth~~ second step?
   a. An intron is removed
   b. A spliceosome forms
   c. A 5′ cap is added
   d. A 3′ poly-A tail is added
   e. Cleavage occurs near AAUAAA

Use this metabolic pathway to answer questions 9-11.

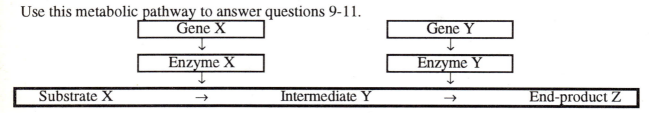

9. If gene X is inducible, it will be transcribed in the
   a. Presence of substrate X.
   b. Presence of intermediate Y.
   c. Presence of end-product Z.
   d. Absence of substrate X.
   e. Absence of end-product Z.

10. If gene X is repressible, it will be transcribed in the
    a. Presence of substrate X.
    b. Presence of intermediate Y.
    c. Presence of end-product Z.
    d. Absence of substrate X.
    e. Absence of end-product Z.

11. If end-product Z acts as a co-repressor, this represents
    a. An inducible operon.
    b. A repressible operon.
    c. A supercoiled operon.
    d. An attenuated operon.
    e. None of the above.

12. One preparation of single-stranded DNA has a Cot value of $10^{-3}$ and the Cot value of another preparation is $10^3$. The faster reassociation time indicates the presence of
    a. Unique-sequence DNA.
    b. Exons.
    c. Repeated DNA sequences.
    d. Introns.
    e. mRNA.

13. If an inactive nucleus from a tadpole intestinal cell is transplanted into an enucleated frog egg, the nucleus will become active and the egg will develop into a frog. This shows
    a. Absences of transcription factors.
    b. Attenuation.
    c. Cytoplasmic control.
    d. Gene amplification.
    e. Methylation.

14. Which one of the following does **not** belong in this list?
    a. Acetylated histones (H3 and H4)
    b. Histone H1
    c. High HMG-protein content
    d. Attachment to the nuclear matrix
    e. Z- DNA

Use this figure to answer questions 15-16.

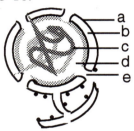

15. The perinuclear space.

16. The nucleoplasm.

Use this figure to answer question 17.

| lacI | ............ | Promoter | Operator | lacZ | lacY | lacA |

17. In the presence of lactose and glucose,
    a. CAP binds cAMP; glucose binds the repressor—transcription is inhibited.
    b. CAP binds the operator; glucose binds the repressor—transcription is active.
    c. CAP is inactive; lactose binds the repressor—transcription is inhibited.
    d. CAP-cAMP bind to the promoter; lactose binds the repressor—transcription is active.
    e. Glucose binds the operator to inhibit transcription.

18. When ecdysone circulates in the blood of an insect all of the following happen to initiate molting **except**
    a. Ecdysone enters the nucleus.
    b. The dimer binds to DNA via zinc fingers.
    c. The steroid binds to a cell-surface receptor.
    d. Transcription begins.
    e. Two hormone-receptor complexes form a dimer.

# Chapter 10

19. Most of the hnRNP in the nucleus is
    a. Introns.
    b. mRNA.
    c. pre-RNA.
    d. rRNA.
    e. tRNA.

20. Which of the following is **not** normally located upstream from the promoter?
    a. Preinitiation complex
    b. Response element
    c. RNA polymerase binding
    d. Structural genes
    e. Enhancer sequence

21. All of the following are true about the polytene chromosome in the salivary gland of *Drosophila* **except**
    a. It displays puffs where transcription is occurring.
    b. It is made of about 1000 strand of DNA.
    c. It is transcriptionally inactive.
    d. Nonhistone proteins accumulate at the puffs.
    e. Several genes are located in each band

# PROBLEMS

1. Locate the promoter, the structural gene, and the termination site. Is this prokaryotic or eukaryotic DNA?

```
     1
5'  T A T A A T T A C C G G G C C T A T G A A A G T C A
3'  A T A T T A A T G G C C C G G A T A C T T T C A G T
    27
5'  T A A T G G C C C T A A C C C C C C G G G G G G
3'  A T T A C C G G G A T T G G G G G G C C C C C C
```

2. If *lacZ* is inserted into one of the X chromosomes of an embryonic cell, the cell produces β-galactoside which can be easily detected. If daughter cells are separated and cultured, about half the cells produce β-galactosidase and the others do not. Explain why all the cells do not produce the enzyme.

3. *Escherichia coli* bacteria are grown in four different media:

   | Flask | Carbohydrate | Color after ONPG |
   |-------|--------------|------------------|
   | 1 | None | None |
   | 2 | 1.0% lactose | Yellow |
   | 3 | 1.0% glucose | None |
   | 4 | 1.0% glucose and 1.0% lactose | None |

   After 1 hour, the cells are killed and ortho-nitrophenyl galactoside (ONPG) is added to each flask. The enzyme β-galactosidase will hydrolyze ONPG to ortho-nitrophenol (bright yellow) and galactose. What can you conclude from these data?

4. The plant growth hormone phytochrome influences many aspects of plant development including flowering. The presence of phytochrome causes an increase in amylase and inhibitors of transcription stop amylase production. Research is currently going on to

## Chapter 10

determine the method of action of phytochrome. Phytochrome activates the second messenger DAG. How could this activate amylase production? If phytochrome is found in the nucleus of the cell, it will suggest an alternative method of action. Describe a likely method of action in the nucleus.

5. Most *ras* proto-oncogene mRNA contains an extra exon that includes a number of UAA codons. Cancer cells produce *ras* mRNA missing this exon. Why doesn't the longer mRNA cause cancer? Where does the mistake most likely occur in the abnormal *ras* product?

6. *Photobacterium* bacteria have a *lux* operon that consists of a promoter, operator, and structural genes for the enzyme luciferase. Luciferase binds with $FMNH_2$ and emits a photon of light to oxidize the $FMNH_2$. *E. coli* bacteria have an *ara* operon. When arabinose binds to the repressor, the structural genes are transcribed to degrade arabinose. Assume that you engineer *Pseudomonas* bacteria to contain the *ara* repressor gene and the *lux* structural genes. What happens when the recombinant *Pseudomonas* is exposed to (a) *lux* inducer, (b) arabinose, and (c) lactose.

7. The *fos* and *jun* genes make transcription factors but only in the presence of PDGF. Fos and Jun proteins are present in large amounts in cancer cells but their genes are not considered oncogenes. What gene might be the actual oncogene? PDGF is not found in the cytoplasm or nucleus of target cells. Suggest a mechanism by which PDGF can stimulate Fos and Jun production. Why are both Fos and Jun needed?

## ANSWERS TO STUDY QUESTIONS

| | | | | | | | |
|---|---|---|---|---|---|---|---|
| 1. | a | 2. | c | 3. | a | 4. | b |
| 5. | c | 6. | c | 7. | d | 8. | b |
| 9. | a | 10. | e | 11. | b | 12. | c |
| 13. | c | 14. | b | 15. | b | 16. | d |
| 17. | c | 18. | c | 19. | c | 20. | d |
| 21. | c | | | | | | |

# CHAPTER 11

# *The Ribosome and Translation of Genetic Information*

## ⇨ LEARNING OBJECTIVES

Be able to
1. Describe the physical and chemical structure of a ribosome. Differentiate between prokaryotic and eukaryotic ribosomes.
2. Describe five methods used to discover the molecular structure of ribosomes.
3. Describe the steps in synthesis of ribosomes. List the steps in processing pre-rRNA.
4. Define the following: ribozyme, the wobble hypothesis, antisense RNA.
5. List the steps of protein synthesis.
6. List four ways in which translation can be controlled.
7. Describe how proteins made on free ribosomes get into each of the following:
   a. The nucleus
   b. The mitochondrion
   d. The chloroplast
   e. The peroxisome

## ⇨ CHAPTER OVERVIEW

I. **INTRODUCTION, p. 470**
   1. The role of the ribosome in protein synthesis resembles that of an enzyme.
   2. Unlike an enzyme, a ribosome contains more than 50 different proteins and several kinds of RNA.

II. **RIBOSOME STRUCTURE, pp. 470-476**

   A. **The Use of Electron Microscopy and Subcellular Fractionation Led to the Discovery of the Ribosome and Its Role in Protein Synthesis, pp. 470-471**
      1. Palade *et al.* defined ribosomes after finding them in the microsomal fraction.
      2. The diameter of a ribosome is ~25 nm.
      3. Ribosomes are the site of protein synthesis.

   B. **Ribosomes Are Constructed from Large and Small Subunits, pp. 471-472**
      1. 70S prokaryotic ribosomes consists of a 50S and 30S subunit.
      2. 80S eukaryotic ribosomes consist of a 60S and 40S subunit.
      3. The small subunit has a platform, cleft, head, and base; the large subunit has a central protuberance, ridge, and stalk.
      4. The subunits consist of rRNA and ribosomal proteins:

|          | Prokaryotic (70S) |         | Eukaryotic (80S) |                      |
|----------|-------------------|---------|------------------|----------------------|
|          | 30S               | 50S     | 40S              | 60S                  |
| rRNA     | 16S (1)           | 23S, 5S | 18S (1)          | 28S, 5S, 5.8S (1 each) |
| Proteins | 21 S              | 32 L    | ~33              | ~49                  |

145

# Chapter 11

    C. **Ribosomes Can Be Reconstituted from Mixtures of Purified Ribosomal Proteins and RNA, p. 473**
        1. Reconstitution experiments have shown that ribosomes are assembled in a stepwise fashion.
        2. Ribosomes from mutant cells have been used to locate the targets of inhibitors of protein synthesis.

    D. **The Molecular Structure of Ribosomes Has Been Investigated Using a Variety of Chemical, Physical, and Microscopic Techniques, pp. 473-475**
        1. Crosslinking agents link neighboring proteins and can be used to identify proteins that are next to each other.
        2. The locations of ribosomal proteins made with $^2$H can be detected by neutron diffraction.
        3. Ribonuclease digestion of ribosomes and radioactive labeling identify the sites of RNA-protein interaction.
        4. Antibodies against individual ribosomal proteins permit the surface of the ribosome to be mapped.

    E. **Models of the Bacterial Ribosome Reveal Where Specific Events Associated with Protein Synthesis Take Place, pp. 475-476**
        1. The small subunit binds mRNA and tRNA; the large subunit catalyzes peptide bond formation and hydrolyzes GTP.

III. **RIBOSOME BIOGENESIS, pp. 476-487**

    A. **The Nucleolus Is the Site of Ribosome Formation in Eukaryotic Cells, pp. 476-477**
        1. The nucleolus is a membrane-free organelle consisting of granular components, fibrillar centers, and dense fibrillar components.

    B. **The Genes Coding for Ribosomal RNAs Are Extensively Amplified, pp. 477-479**
        1. Eukaryotic cells may have several thousand copies of the genes for 28S, 18S, and 5.8S rRNA and up to 50,000 copies of the 5S rRNA genes.
        2. The amplified rRNA genes may exist as circular DNA molecules separate from the chromosome.

    C. **The Genes for 18S, 5.8S, and 28S rRNAs are Transcribed in the Nucleolar Fibrillar Center into a Single Pre-rRNA Molecule, pp. 479-480**
        1. The fibrillar center consists of a long central fiber (of DNA) and lateral fibrils of rRNA molecules being transcribed.
        2. The three genes are adjacent to one another on the DNA separated by a nontranscribed spacer DNA.
        3. Each group of three genes (a transcription unit) is transcribed as a single pre-rRNA.
        4. Transcribed spacers are removed during processing of pre-rRNA.

    D. **Pre-rRNA Undergoes Several Cleavage Steps During Its Conversion to Mature Ribosomal RNAs, pp. 480-482**
        1. 25-50% of pre-rRNA is removed; pre-rRNA is cleaved by snRNPs and methylated.

# Chapter 11

- E. **Some Ribosomal RNA Precursors Contain Introns That Are Removed by Self-Splicing, p. 482**
  1. Pre-rRNAs that catalyze the removal of introns are called ribozymes.

- F. **The Formation of Ribosomal Subunits Requires 5S rRNA and Ribosomal Proteins in Addition to 18S, 5.8S, and 28S rRNAs, pp. 482-485**
  1. The 5S RNA gene is located outside of the nucleolus; it is transcribed by RNA polymerase II (and TFIIIA).
  2. Forming pre-rRNA associates with proteins before transcription is complete.
  3. Pre-rRNA is synthesized in the fibrillar centers and processed in the dense fibrillar components.

- G. **Prokaryotic Cells Manufacture Ribosomes without a Nucleolus, pp. 485-487**
  1. Prokaryotic cells have a few copies of each rRNA gene.
  2. The 16S, 23S, and 5S genes are transcribed as a unit. They are separated by RNase III later.
  3. Another RNase cleaves pre-rRNA to produce tRNAs in addition to rRNAs.
  4. Bacterial rRNA is methylated and associated with ribosomal proteins during processing.

IV. **PROTEIN SYNTHESIS, pp. 487-503**

- A. **Protein Synthesis Can Be Studied in Cell-Free Systems, p. 487**
  1. Cell-free protein synthesis can be observed *in vitro* if the microsomal, mitochondrial, and cytosol fractions are mixed.
  2. GTP and ATP can replace the mitochondrial fraction.
  3. The cytosol supplies tRNAs and aminoacyl-tRNA synthetases.
  4. The microsomal fraction can be replaced by membrane-free ribosomes.

- B. **Protein Synthesis Proceeds from the N-Terminus to the C-Terminus, pp. 487-488**

- C. **Translation of Messenger RNA Proceeds in the $5' \rightarrow 3'$ Direction, p. 488**

- D. **Ribosomal Subunits Associate and Dissociate During Protein Synthesis, pp. 488-489**

- E. **The Initiation of Protein Synthesis Requires an Initiator Transfer RNA, pp. 489-490**
  1. mRNA usually has an untranslated leader at the 5′ end and a trailer between the coding sequence and the poly-A tail.
  2. Amino acids are attached to the tRNAs by aminoacyl-tRNA synthetases to form aminoacyl-tRNAs.
  3. Bacteria have a tRNA$^{MET}$ and tRNA$^{fMET}$.
  4. tRNA$^{fMET}$ is formylated after joining tRNA; it is an initiator tRNA.
  5. Eukaryotes also have two tRNAs carrying methionine, one is an initiator tRNA; neither is formylated.

# Chapter 11

F. **During Initiation mRNA Binds to the Small Ribosomal Subunit and the AUG Start Codon Becomes Properly Aligned, pp. 490-491**
 1. In prokaryotes, the Shine-Dalgarno sequence is 5-10 bases prior to the AUG start codon.
 2. The Shine-Dalgarno sequence is complementary to the 3′ end of the 16S rRNA.
 3. In eukaryotes, mRNA binds by the 5′ cap.
 4. Translation is initiated at the first AUG following the 5′ cap.

G. **Initiation Factors Promote the Interaction Between Ribosomal Subunits, mRNA, and Initiator tRNA, pp. 491-492**
 1. In prokaryotes, three initiation factors are required.
    a. IF-1, IF-3, and IF-2/GTP associate with the 30S subunit.
    b. The 5′ end of mRNA binds to the 30S subunit.
    c. IF-2 binds the initiator tRNA to the 30S subunit.
    d. IF-3 is released and the complex is the 30S preinitiation complex.
    e. The 50S subunit binds to the 30S preinitiation complex and IF-1 and IF-2 are released with the hydrolysis of GTP by L7/L12.
    f. This forms the 70S initiation complex.
 2. In eukaryotes, 6 groups of initiation factors are required.
    a. eIF-1 through eIF-3 are comparable to IF-1 through IF-3.
    b. eIF-4 recognizes the 5′ cap and hydrolyzes ATP.

H. **Ribosomes Contain a P Site That Binds the Growing Polypeptide Chain and an A Site That Binds Incoming Aminoacyl-tRNAs, pp. 492-494**
 1. During the elongation phase:
    a. Aminoacyl-tRNA binds the next codon after AUG at the A site.
    b. The previously bound tRNA is in the P site.
    c. Binding between the mRNA codons and tRNA anticodons occurs in the cleft between the platform and head of the small unit.
 2. Binding of the new tRNA to the A site requires GTP bound to elongation factor.
 3. Finding the correct aminoacyl-tRNA is a trial-and-error process.
 4. mRNA and tRNA align to allow flexibility at the third base (wobble hypothesis).

I. **Peptide Bond Formation is Catalyzed by Ribosomal RNA, pp. 494-495**
 1. Energy for peptide bond formation comes from the high-energy bond that joins the amino acid to the tRNA at the P site.

J. **Translocation Advances the Messenger RNA by Three Nucleotides, pp. 495-496**
 1. After peptide bond formation, the P site contains an empty tRNA and the A site contains a peptidyl-tRNA.
 2. The mRNA advances by one codon.
 3. The peptidyl-tRNA moves to the P site and the empty tRNA is ejected.
 4. Translocation requires GTP.

K. **Protein Synthesis Is Terminated by Release Factors That Recognize Stop Codons, p. 496**
   1. In prokaryotes, three release factors terminate protein synthesis.
   2. In eukaryotes, one release factor recognizes all three stop codons.
   3. A suppressor mutation allows an alternative amino acid to be placed in a polypeptide if a mutation has caused a stop codon in the middle of the polypeptide chain.

L. **Polysomes Consist of Clusters of Ribosomes Simultaneously Translating the Same Messenger RNA, pp. 497-499**
   1. In prokaryotes, ribosomes can attach to mRNA that is still being transcribed.

M. **Inhibitors of Protein Synthesis Are Useful Tools for Studying Translation, pp. 500-502**
   1. Puromycin binds to the A site.
   2. Streptomycin binds to the small subunit in prokaryotes.
   3. Diphtheria toxin inhibits eEF-2.

N. **Newly Synthesized Polypeptide Chains Are Often Processed by Post-Translational Cleavage Reactions, p. 502**
   1. mRNA codes for extra amino acids creating proinsulin which is cleaved to produce insulin.
   2. Cleavage of a polyprotein produces more than one protein.

O. **The Folding of Newly Synthesized Proteins Is Facilitated by Enzymes and Chaperones, pp. 502-503**
   1. Enzymes facilitate disulfide bond formation.
   2. Chaperones bind to polypeptides and provide stability so incorrect folding doesn't occur.

V. **TRANSLATIONAL AND POST-TRANSLATIONAL CONTROL, pp. 503-511**

   A. **Ribosome Production Is Controlled in Several Different Ways, pp. 503-504**
      1. Protein synthesis is proportional to the number of ribosomes.
      2. In prokaryotes, the stringent response inhibits rRNA and tRNA synthesis.

   B. **Translational Repressors Inhibit Protein Synthesis by Binding to Specific Messenger RNAs, pp. 504-505**
      1. In prokaryotes, ribosomal protein operons code for proteins that function as a ribosomal protein and as a translational repressor.
      2. When [mRNA] > [rRNA], these repressors bind mRNA.
      3. In eukaryotes, ferritin mRNA contains an iron-responsive element (IRE). IRE is bound by IRE-binding protein when iron is absent.
      4. Inactive mRNA (in unfertilized eggs) is bound with proteins to form mRNPs.

   C. **Translation Can Be Regulated by Controlling Messenger RNA Lifespan, pp. 505-506**
      1. For some proteins, the lifespan of mRNA is proportional to the length of the poly-A tail.

# Chapter 11

**D. Antisense RNA Molecules Inhibit the Translation of Complementary Messenger RNAs, pp. 506-507**
   1. Antisense RNA binds to mRNA. Double-stranded RNA does not bind to ribosomes and is susceptible to degradation.

**E. Translation Rates Can Be Increased or Decreased by the Phosphorylation of Protein Synthesis Factors, p. 507**

**F. The Availability of Transfer RNAs Can Influence the Translation of Messenger RNA, pp. 507-508**

**G. Attenuation Allows Translation Rates to Control the Termination of Transcription, p. 508**
   1. Premature termination of transcription is called attenuation.
   2. In the *trp* operon of *E. coli*, mRNA is transcribed and translation begins.
      a. If tRNA$^{TRP}$ is not available when translation reaches the Trp codons, translation stalls. The mRNA folds to permit transcription.
      b. If tRNA$^{TRP}$ is available when translation reaches the Trp codon, mRNA folds to stop transcription.

**H. A Protein's Biological Activity Is Subject to Control by Post-Translational Alterations in Protein Structure and Conformation, p. 509**
   1. Post-translational control includes allosteric inhibition and activation, association and dissociation of polypeptide subunits, acetylation, and phosphorylation.

**I. Differences in Protein Lifespan Are an Important Factor in the Control of Gene Expression, pp. 509-510**
   1. The half-life of a protein is the amount of time required for half of the molecules existing at any time to be degraded.

**J. Ubiquitin Targets Selected Proteins for Degradation, pp. 510-511**
   1. Proteins joined to ubiquitin are targeted for destruction by 26S proteasome.

**VI. PROTEIN TARGETING, pp. 511-515**
   1. Proteins manufactured in the cytoplasm are routed to intracellular and extracellular destinations.
   2. A signal sequence on the forming polypeptide causes the ribosome to attach to the ER. These proteins will go to the Golgi complex, lysosomes, plasma membrane, or be secreted.
   3. Free ribosomes make proteins destined for the cytosol, nucleus, mitochondria, chloroplasts, or peroxisomes.

**A. Proteins Are Directed into the Nucleus by a Bipartite Nuclear Localization Sequence Enriched in Basic Amino Acids, pp. 511-512**
   1. A nuclear localization sequence (of amino acids) targets a protein for uptake by the nucleus.
   2. Some proteins transported into the nucleus are bound to RNA.

### B. Targeting Sequences Direct Mitochondrial Proteins Made in the Cytosol to their Appropriate Locations within the Mitochondrion, pp. 512-514

1. Proteins synthesized in the cytosol as precursor molecules containing a matrix-targeting signal are transported into the mitochondrial matrix. The outer membrane receptor becomes associated with the inner membrane to form a translocation channel. A membrane potential is necessary for translocation.
2. Proteins targeted to the inner mitochondrial membrane are transported into the matrix and then inserted into the membrane
3. Proteins destined for the inner membrane space may contain a matrix-targeting signal.
4. Proteins for the outer membrane contain a matrix-targeting signal and a hydrophobic amino acid chain which causes them to be retained in the membrane.

### C. Targeting Sequences Direct Chloroplast Proteins Made in the Cytosol to their Appropriate Locations within the Chloroplast, p. 514

1. Proteins destined for the stroma contain a stromal-targeting signal. Translocation occurs at contact points between the inner and outer membranes. ATP is required for translocation.
2. Thylakoid-lumen proteins are transported into the stroma and then moved into the thylakoid lumen.
3. Thylakoid-membrane proteins are transported into the stroma and then inserted into the membrane.

### D. Targeting Sequences Guide Proteins Made in the Cytosol into Peroxisomes, Glyoxysomes, and Glycosomes, pp. 514-515

1. Peroxisomal proteins are synthesized on free ribosomes; a targeting signal causes their incorporation into peroxisomes.

## KEY TERMS

26S proteasome
antisense RNA
chaperone
elongation factor
initiation factor
iron-responsive element
nucleolus
polyprotein
ribozyme
Shine-Dalgarno sequence
suppressor mutation
translational control
ubiquitin

A (aminoacyl) site
anuclear mutant
crosslinking agents
expression vector
initiator tRNA
mRNPs
P (peptidyl) site
puromycin
rolling circle replication
streptomycin
transcription unit
translational repressor
wobble hypothesis

affinity labels
attenuation
diphtheria toxin
half-life
IRE-binding protein
nuclear localization sequence
peptidyl transferase
ribosome
rRNA
stringent response
transferrin receptor
tRNAs

## KEY FIGURE

Label the following:
30S subunit
50S subunit
A site
Aminoacyl-tRNA binding
mRNA
P site
Peptide bond formation
Translocation
tRNA

Indicate where GTP is required. Is this a prokaryote or eukaryote? Identify the second and third amino acids that are encoded by the mRNA shown.

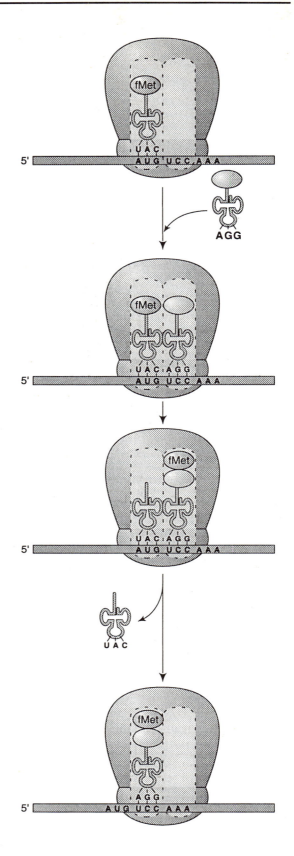

# Chapter 11

# ⇪ STUDY QUESTIONS

1. Following subcellular fractionation and differential centrifugation, ribosomes are found in the
   a. Nuclear fraction.
   b. Mitochondrial fraction.
   c. Microsomal fraction.
   d. Ribosomal fraction.
   e. Cytosol fraction.

2. Eukaryotic ribosomes consist of
   a. A 70S unit.
   b. 40S and 60S subunits.
   c. 30S and 50S subunits.
   d. A 100S unit.
   e. None of the above.

3. Which one of the following is **not** part of a prokaryotic ribosome?
   a. A 30S subunit consisting of 16S rRNA
   b. A 50S subunit consisting of 23S and 5S rRNA
   c. A 30S subunit consisting of 21 proteins
   d. A 40S subunit consisting of 18S rRNA
   e. None of the above

4. A eukaryotic cell has
   a. One copy of each ribosomal transcription unit.
   b. Thousands of copies of the 5S rRNA gene.
   c. Two copies of each gene for rRNA and ribosomal proteins.
   d. Two copies of each gene for rRNA and one copy of each ribosomal protein gene.
   e. None of the above.

5. The nucleolus
   a. Is the site of protein synthesis.
   b. Is the site of pre-rRNA synthesis.
   c. Is surrounded by a membrane.
   d. Is near the nucleus.
   e. All of the above.

6. Which one of the following is **incorrect** for the biogenesis of eukaryotic ribosomes?
   a. 25-50% of the pre-rRNA is degraded
   b. Pre-rRNA binds to ribosomal protein
   c. Pre-rRNA is cut to form three classes of rRNA
   d. Pre-rRNA is methylated
   e. The rRNA gene for the 40S subunit is transcribed separately

7. If cells are given radioactive amino acids and then lysed and the cellular components separated by centrifugation, radioactive proteins are detected in the
   a. Nuclear fraction.
   b. Mitochondrial fraction.
   c. Microsomal fraction.
   d. Ribosomal fraction.
   e. All of the above.

8. Cell-free protein synthesis can be observed if the microsomal, cytosol, and mitochondrial fractions are mixed together. The cytosol fraction can be replaced with
   a. tRNAs and aminoacyl-tRNA synthetases.
   b. Amino acids.
   c. GTP and ATP.
   d. NADH.
   e. None of the above because cytosol is required.

9. The following steps occur to initiate protein synthesis. What is the third step?
   a. GTP is hydrolyzed by the large subunit.
   b. IF-1, IF-3, and IF-2/GTP associate with the 30S subunit.
   c. IF-2 binds the initiator tRNA to the 30S subunit.
   d. The 5′ end of mRNA binds to the 30S subunit.
   e. The 50S subunit binds.

10. Which of the following does **not** represent complementary base-pairing?
    a. Shine-Dalgarno sequence and 16S rRNA
    b. AUG and tRNA$^{fMET}$
    c. mRNA 5′ cap and the 60S subunit
    d. AUG and tRNA$^{Met}$
    e. None of the above

11. The following steps occur during the elongation phase. The mRNA sequence is AUGUUU. Which is the third step?
    a. A peptide bond is formed.
    b. tRNA$^{AAG}$ binds to the A site.
    c. tRNA$^{AAG}$ is in the P site.
    d. tRNA$^{GUU}$ enters the A site, then leaves.
    e. tRNA$^{UAC}$ is in the P site.

12. Which of these does **not** inhibit translation?
    a. Proteins that bind mRNA if [rRNA] is low.
    b. Proteins that bind mRNA if substrate for the encoded protein is low.
    c. Formation of mRNP complexes.
    d. Increased production of ribosomes.
    e. None of the above.

13. All of the following are true about antisense RNA **except** it
    a. Is complementary to mRNA.
    b. Forms double-stranded RNA.
    c. Joins with the 30S subunit.
    d. Inhibits protein synthesis.
    e. All of the above.

14. The following events occur in *E. coli* involving the *trp* operon if tRNA$^{TRP}$ is not available. What is the third step?
    a. mRNA continues to be transcribed.
    b. mRNA is transcribed.
    c. The mRNA folds.
    d. Translation begins.
    e. Translation stalls.

# Chapter 11

15. Protein synthesis can be regulated by all of the following **except**
    a. Allosteric inhibition.
    b. Availability of tRNA.
    c. Degrading mRNA.
    d. Phosphorylation of elongation factors.
    e. None of the above.

16. Which one of the following is a post-translation event?
    a. Allosteric inhibition
    b. Attenuation
    c. mRNA binding proteins
    d. Inhibition of tRNA synthesis
    e. None of the above

17. Proteins synthesized on free ribosomes can be found in all of the following **except**
    a. Nucleus.
    b. Mitochondria.
    c. Plasma membrane.
    d. Chloroplasts.
    e. Peroxisomes.

18. The antibiotic kasugamycin blocks binding of tRNA$^{FMET}$. From this information you can conclude that kasugamycin
    a. Prevents polypeptide termination in eukaryotes.
    b. Prevents polypeptide elongation in eukaryotes.
    c. Prevents polypeptide initiation in prokaryotes.
    d. Prevents mRNA—ribosome binding in eukaryotes.
    e. Prevents mRNA—ribosome binding in prokaryotes.

19. A preinitiation complex consists of all of the following **except**
    a. A 30S subunit
    b. mRNA
    c. A 50S subunit
    d. IF-1
    e. GTP

20. Mitochondrial matrix proteins differ from chloroplast stromal proteins in that the chloroplast proteins
    a. Are synthesized in the cytosol.
    b. Are synthesized by ribosomes on the ER.
    c. Cross the chloroplast envelope at translocation channels.
    d. Require ATP to enter the chloroplast.
    e. All of the above.

21. Which of the following **leads** to all the others?
    a. If-3 is not released from the 30S subunit
    b. mRNA doesn't bind to the 30S subunit
    c. The 50S subunit doesn't bind to the 30S subunit
    d. tRNA$^{FMET}$ doesn't bind to the 30S subunit
    e. The 70S initiation complex doesn't form

## PROBLEMS

1. Scherrer and Darnell incubated cells for a short time with $^3$H-uridine and then added actinomycin D which binds GC sequences in DNA. They analyzed the homogenate at time intervals for radioactive RNAs. Of all the molecules in the cell, how did they know that the radioactive molecules would be RNA? What was the purpose of the actinomycin D? Their results are shown below. Use these data to explain the formation of rRNAs.

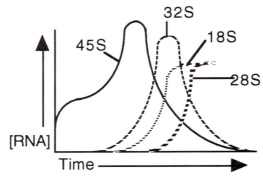

2. Members of the Picornaviridae family include *Rhinovirus* species (common cold) and *Enterovirus* species (e.g., polio viruses). These viruses lack a 5' cap and produce an enzyme that destroys eIF-4. Of what advantage is this enzyme to the virus?

3. The single-stranded viral genome of the Picornaviridae is approximately 7,500 bases encoding four capsid proteins. A Western blot technique can be used to detect these proteins in an infected cell. The results are given below. Describe the synthesis of these viral proteins.

| Lane | Contains | Viral proteins |
|---|---|---|
| 1 | Control cells | None |
| 2 | Infected cells treated with actinomycin D 5 minutes after infection | None |
| 3 | Infected cells 10 minutes after infection | 33 kd |
| 4 | Infected cells 20 minutes after infection | 30, 28, 25, 8 kd |
| 5 | Infected cells treated with a protease inhibitor 5 minutes after infection | 33 |
| 6 | Infected cells treated with a protease inhibitor 10 minutes aftrer infection | 63 |
| 7 | Infected cells treated with a protease inhibitor 20 minutes after infection | 91 |

4. Some tomatoes have a very short shelf-life because they produce large amounts of the ripening hormone ethylene. Ethylene is a small molecule containing only two carbon atoms. It is a gas at room temperature. Explain how antisense technology has been employed to lengthen the shelf-life of these tomatoes.

5. Some plants produce a toxin called proteinase inhibitor initiation factor (PIIF) that inhibits protein digestion in insects. PIIF may be a small polypeptide. Provide a possible mechanism of action for this toxin.

6. The normal targeting sequence for peroxisomal enzymes is rich in serine and threonine and ends in SKL. In humans with the lethal autosomal recessive disorder called primary

## Chapter 11

hyperoxaluria I, the SKL is omitted from aminotransferase I. The enzyme is properly transcribed and translated. Why, then, do these people have a lethal disorder?

7. Degeneration of muscle tissues can be induced by starvation or, in the case of insect metamorphosis, can be a programmed part of the life cycle. In both cases, the ubiquitin gene is induced and ubiquitin can be detected in these cells using florescent-labeled antibodies. Explain the significance of the increase in ubiquitin in these cells.

## ANSWERS TO STUDY QUESTIONS

| 1. c | 2. b | 3. d | 4. b |
|------|------|------|------|
| 5. b | 6. e | 7. c | 8. a |
| 9. c | 10. c | 11. b | 12. d |
| 13. c | 14. e | 15. a | 16. a |
| 17. c | 18. c | 19. c | 20. d |
| 21. b | | | |

# CHAPTER 12

# Cell Cycles and Cell Division

## ⇧ LEARNING OBJECTIVES

Be able to
1. Define the following terms:
    a. Cell cycle
    b. Synchronized population
    c. Temperature-sensitive mutant
2. Describe binary fission.
3. Differentiate between
    a. Binary fission and asymmetric fission
    b. ppGpp and *oriC*
    c. Binary fission and endospore formation
    d. Chromosome and chromatid
    e. Centromere and centrosome
    f. Kinetochore and centromere
    g. Mitosis and cytokinesis
    h. Cytokinesis in plants and animals
    i. Mitosis and meiosis
    j. Meiosis I and meiosis II
    k. Interphase and mitosis
    l. Anaphase I and anaphase II
4. Describe the events in each of the following:
    a. $G_1$ phase
    b. S phase
    c. $G_2$ phase
5. Explain the role of each of the following in the eukaryotic cell cycle: cdc2 kinase, MPF, and *cdc25* gene.
6. Identify the role of each of the following in mitosis:
    a. ARS element
    b. *CEN* sequence
    c. Tel sequence
    d. Replicon
    e. H1
7. Describe the events of prophase, metaphase, anaphase, and telophase in mitosis.
8. Describe the five stages of prophase I.
9. Define each of the following terms:
    a. Homologous recombination
    b. Conjugation
    c. Transduction
    d. Transformation

## ⇧ CHAPTER OVERVIEW

### I. STUDYING THE CELL CYCLE, pp. 518-519
   1. The cell cycle consists of DNA replication, mitosis, and cytokinesis.

   **A. Synchronized Cell Populations Can Be Produced Using Either Induction or Selection Techniques, pp. 518-519**
   1. Synchronized populations are in the same stage of the cell cycle at any given moment.
   2. When all cells are blocked in one stage of division, the blocking agent can be removed to get induction synchrony.
   3. Cells in a particular stage in the cell cycle can be removed from a culture in selection synchrony.

# Chapter 12

- B. **Conditional Mutants Are Useful for Identifying Important Events in the Cell Cycle, p. 519**
    1. Temperature-sensitive mutants can be grown normally at one temperature but they exhibit an abnormal trait at a different temperature.

II. **BACTERIAL DIVISION, pp. 519-520**

- A. **The Prokaryotic Cell Cycle Is Constructed from C and D Periods That Can Overlap and an Optional B Period, p. 519**
    1. The Cooper-Helmstetter model states that the prokaryotic cell cycle includes the C period (time required for replication of the chromosome) and the D period (time between the end of DNA replication and the completion of cell division).
    2. In populations that divide once per hour: C = 40 min. and D = 20 min.
    3. In populations that divide more slowly, the B period occurs between cell division and the initiation of DNA synthesis.
    4. If a new round of DNA replication starts before completion of the previous round, cell cycles shorter than 60 min. can be created.

- B. **Changing Concentrations of ppGpp May Link Growth Rate to the Initiation of DNA Replication, pp. 519-520**
    1. To maintain the proper size, cell division must be coordinated to cell growth; and the growth rate is usually dependent on the availability of nutrients.
    2. ppGpp accumulates in slow growing cells and depresses transcription of genes required to initiate DNA synthesis.

- C. **Bacterial Chromosome Replication is Bidirectional Starting from a Single Origin, pp. 520-521**
    1. The initiation of *E. coli* chromosome replication requires a 245 base-pair stretch of DNA called *oriC*.
    2. DnaA protein binds to *oriC* in an ATP-dependent reaction, separating the two DNA strands for replication.

- D. **Septum Formation Leads to Chromosome Segregation and Cell Division, pp. 521-522**
    1. When a cell has doubled in size, it divides symmetrically by binary fission.
    2. A few prokaryotes undergo asymmetrical fission.
    3. When FtsZ protein accumulates around the cell's equator, the plasma membrane grows inward to produce a septum.
    4. The chromosome is attached to the plasma membrane in at least two places and as the membrane grows inward, new membrane is added to separate the newly forming DNA molecules to each end of the cell.
    5. A thick cross wall forms between the two membranes of the septum; the cross wall is cleaved to separate the cells.
    6. Bacteria grown in the presence of penicillin produce protoplasts.

- E. **Sporulation Is a Special Type of Cell Division That Converts Bacteria into Dormant Spores, p. 522**
    1. Some gram-positive bacteria produce an endospore to survive unfavorable conditions.
    2. The endospore contains a chromosome and is surrounded by a highly resistant spore coat.

# Chapter 12

3. In favorable conditions, the endospore will germinate to produce a normal cell.

## III. MITOTIC DIVISION, pp. 522-546

### A. The Eukaryotic Cell Cycle Is Divided into Four Phases Called $G_1$, S, $G_2$, and M, pp. 522-523
1. Interphase consists of $G_1$, S, and $G_2$.
2. During M, mitosis (nuclear division) and cytokinesis (cell division) occur.
3. M is followed by $G_1$, then the S phase during which chromosomes are replicated, then $G_2$.

### B. Progression through the Cell Cycle Is Controlled by Cyclin-Dependent Protein Kinases, pp. 523-526
1. Lack of nutrients causes the cell cycle to stop in $G_1$ at the restriction point or START; cells are said to be in the $G_0$ state.
2. Passing beyond START commits a cell to division.
3. cdc2 kinase triggers the transition from $G_1$ to S.
4. Maturation promoting factor (MPF) triggers the transition from $G_2$ to M.
5. Cells continuously make cyclins. The concentration increases during interphase.
    a. $G_1$ cyclin combines with and activates cdc2 kinase. The cell enters S.
    b. Mitotic cyclin combines with and inactivates cdc2 kinase to form MPF. The cell enters $G_2$.
    c. When the *cdc25* gene product activates cdc2 kinase the cell enters M.
    d. Mitotic cyclin is degraded.
    e. Following mitosis, the cells enter $G_1$ and the cyclin concentration begins to increase again.

### C. Cell Division in Multicellular Organisms Requires Cell-Specific Growth Factors, pp. 526-527
1. Cell growth is controlled by extracellular signaling molecules called growth factors.
2. In cell culture, cell growth and division can be triggered by PDGF.
3. Growth factors interact with receptors that function as protein-tyrosine kinases. Binding to a growth factor causes autophosphorylation of the receptor.
4. Then, *Ras* is activated and stimulates mitogen-activated protein (MAP) kinases.
5. MAP kinases activate other proteins including AP1 transcription factor.
6. Plant growth hormones include:
    a. Auxins—promote cell growth
    b. Gibberellins—promote cell growth
    c. Cytokinins—promote cell division in the presence of auxins
    d. Abscisic acid—inhibits growth
    e. Ethylene—inhibits growth

### D. Eukaryotic Chromosome Replication is Semiconservative, p. 527
1. The two new chromosomes remain attached as chromatids until the cell divides.
2. The attachment site is the centromere.

## Chapter 12

- **E. Eukaryotic Chromosomes Are Replicated Using Multiple Replicons, pp. 527-530**
  1. DNA replication is initiated at multiple sites to produce replicons of 50,000 to 300,000 base pairs.
  2. DNA replication is bidirectional from the replication origin.
  3. Autonomously replicating sequence (ARS) elements have the ability to replicate.
  4. Eukaryotic DNA is replicated at ~2,000 base pairs/min.
  5. The duration of the S phase is proportional to the number of replicons and the rate at which they are activated.
  6. DNA synthesis is restricted to a few hundred foci in the nuclear matrix.

- **F. Active and Inactive Genes Are Replicated at Different Times during S Phase, pp. 530-532**
  1. Heterochromatin is replicated late.
  2. Heterochromatin is located near centromeres and telomeres and comprises the bulk of the inactive X chromosome.

- **G. Nucleosome Assembly during S Phase Requires the Synthesis of New Histones, pp. 532-533**
  1. New histones are mostly on the lagging strand.

- **H. Telomeres Protect the Ends of Linear Chromosomes from Being Shortened during Replication, pp. 533-534**
  1. Telomeres are highly repeated satellite DNA at the ends of a linear chromosome.
  2. TTAGGG is a human Tel sequence.
  3. During DNA replication, the 3′ end of the molecule is shortened.
  4. Telomerase makes more copies of the Tel sequence to compensate for the shortening.
  5. In order for a linear DNA molecule to function as a chromosome it must:
     a. Replicate during the S phase (requires ARS element).
     b. Segregate during mitosis (requires *CEN* sequence).
     c. Maintain itself through repeated generations (requires Tel sequence).

- **I. M Phase Is Subdivided into Prophase, Metaphase, Anaphase, Telophase, and Cytokinesis, pp. 534-536**
  1. During prophase, the chromosomes condense, the mitotic spindle forms, and the nuclear envelope breaks down.
  2. During metaphase, chromosomes attach to the spindle and migrate to the equator.
  3. During anaphase, paired chromatids separate and move to opposite poles.
  4. During telophase, chromosomes uncoil into chromatin, the spindle disappears, and the nuclear envelopes form.

- **J. Protein Phosphorylation Initiates Prophase by Triggering Chromosome Condensation and Nuclear Envelope Breakdown, pp. 536-537**
  1. MPF phosphorylates H1 and chromosomes condense.
  2. MPF phosphorylates lamin and then the lamin depolymerizes.

K. **The Mitotic Spindle Is Generated from Microtubules Whose Growth Is Initiated by Centrosomes, p. 537**
   1. Preexisting tubulin polymerizes to make the mitotic spindle.
   2. The spindle forms from the centrosomes; long microtubules form between the centrosomes; short microtubules form asters.
   3. Centrosomes duplicate during the S phase and migrate toward opposite poles of the cells.

L. **At the Beginning of Metaphase, Chromosomes Attach to the Mitotic Spindle through their Kinetochores, pp. 537-539**
   1. The centomere consists of the *CEN* sequence and kinetochore.
   2. Two chromatids are attached at their *CEN* sequences.
   3. Kinetochore microtubules attach to the kinetochores of each paired chromatid.

M. **Metaphase Chromosomes Are Moved to the Spindle Equator by Two Kinds of Forces, pp. 539-540**
   1. Chromatids migrate to the equator because they are pulled by the opposite pole and pushed from the nearest pole.

N. **Metaphase Chromosomes Are Intricately Folded Chromatin Fibers, pp. 540-541**
   1. Karyotyping allows microscopic identification of individual chromosomes.
   2. During metaphase, each chromosome is compacted to ~0.0001x its extended length.
   3. When karyotypes are prepared and stained, each folded chromosome has a unique pattern of G (high [A-T]) and R (high [G-C]) bands.
   4. The folding is due to a protein scaffold (topoisomerase II) attached to the DNA.

O. **Anaphase Is Initiated by the Splitting of Chromosomal Centromeres, pp. 541-542**
   1. Topoisomerase II is one of the components needed to split the centromere.
   2. INCENPs hold paired chromatids together until anaphase.

P. **Chromosomes Move toward the Spindle Poles during Anaphase A, pp. 542-543**
   1. During anaphase A chromosomes are pulled to the poles by depolymerization of kinetochore microtubules.
   2. Kinetochores contain motor proteins that generate movement by advancing along a surface, accompanied by ATP hydrolysis.

Q. **The Spindle Poles Move away from Each Other during Anaphase B, p. 543**
   1. During anaphase B the spindle poles move away from each other.
   2. Polar microtubules lengthen and opposing microtubules slide apart.

R. **During Telophase, Nuclear Envelopes Form around the Two Sets of Chromosomes, pp. 543-544**
   1. Membrane vesicles begin to condense around the uncoiling chromosomes; the nuclear lamina reassembles.

S. **Cytokinesis in Animal Cells Is Mediated by a Contractile Ring That Pinches the Cell in Two, pp. 544-546**
1. The midbody forms at the equator during anaphase.
2. The plasma membrane invaginates forming the cleavage furrow.
3. Actin filaments develop beneath the cleavage furrow to form the contractile ring.

T. **Cytokinesis in Plant Cells Progresses from the Center of the Cell toward the Periphery, p. 546**
1. Prior to mitosis a band of microtubules forms the preprophase band.
2. In late anaphase, vesicles accumulate to form the phragmoplast.
3. As the vesicles grow outward they fuse to generate a double-layered membrane.
4. Cell wall material is deposited forming the cell plate between the two membranes.

IV. **MEIOTIC DIVISION AND GENETIC RECOMBINATION, pp. 546-559**

A. **Meiosis Converts One Diploid Cell into Four Haploid Cells That Are Genetically Different, p. 548**
1. A diploid cell has two copies of each homologous chromosome.
2. Homologous chromosomes are the same type of chromosome.
3. A haploid cell has one of each pair.
4. Haploid cells are generated by meiosis which reduces the number of chromosomes.
5. In meiosis, DNA is replicated and the cell divides twice to produce 4 haploid cells from one diploid cell.
6. Genetic recombination occurs in meiosis to create new gene combinations.

B. **Homologous Chromosomes Become Paired and Exchange DNA during Prophase I, pp. 549-551**
1. DNA is replicated during the S phase. Each chromosome consists of two chromatids entering prophase I.
2. Prophase I includes
   a. Condensation of chromatin (leptotene stage).
   b. Pairing or synapsis of homologous chromosomes (zygotene stage).
   c. Thickening of chromosomes and formation of tetrads consisting of four chromatids (2 chromosomes) (pachytene stage).
   d. Homologous chromosomes begin to separate remaining attached at the chiasmata; chromatids uncoil and nucleoli form (diplotene stage).
   e. Chromosomes recondense, chiasmata migrate to the ends of the chromosomes (terminalization), microtubules begin to form a spindle, and the nuclear envelope disappears (diakinesis).

C. **Tetrads Align at the Spindle Equator at Metaphase I, pp. 551-552**
1. Each of the four chromatids has a kinetochore.
2. Each kinetochore located on the two chromatids of one chromosome attaches to microtubules from the same pole.

D. **Homologous Chromosomes Move to Opposite Spindle Poles during Anaphase I, p. 552**
1. The pair of chromatids making up one chromosome migrate to the same pole.
2. The centromere did not divide as in anaphase of mitosis.

## Chapter 12

E. **Telophase I and Cytokinesis Produce Two Haploid Cells Whose Chromosomes Consist of Paired Chromatids, pp. 552-553**

F. **Meiosis II Produces Gametes by a Process That Resembles a Mitotic Division, p. 553**
1. During meiosis II, microtubules join opposite sides of the centromere that links two chromatids.
2. The centromere divides and the chromatids are separated.

G. **Genetic Recombination Occurs in Both Prokaryotes and Eukaryotes, pp. 553-554**
1. Homologous recombination occurs when two DNA molecules with similar sequences exchange segments.
2. In bacterial conjugation, the chromosome from one bacterium is injected into a recipient bacterium; the DNA in the recipient recombines.
3. In transformation, one bacterium takes up naked DNA which can recombine with DNA in the cell.
4. In transduction, a virus transfers DNA from one bacterium to another.

H. **Breakage-and-Exchange Occurs during Homologous Recombination, pp. 554-555**
1. Chromosomes break and exchange segments in homologous recombination.

I. **Homologous Recombination Can Lead to Gene Conversion, pp. 555-556**
1. Nonreciprocal recombination is called gene conversion (see J.2 below).

J. **Homologous Recombination Involves the Formation of Single-Strand Exchanges (Holliday Junctions), pp. 556-558**
1. Exchange of single strands of DNA between homologous chromosomes (Holliday Junction) results in noncomplementary regions in the chromosome.
2. A cell can remove and repair one of the strands which would lead to gene conversion.
3. A cell could go through mitosis and produce two genetically different cells.

K. **The Synaptonemal Complex Facilitates Chromosome Pairing and Recombination during Meiosis, pp. 558-559**
1. The synaptonemal complex is a protein-containing structure that joins homologous chromosomes.
2. It forms during the leptotene stage.

## ☞ KEY TERMS

| | | |
|---|---|---|
| anaphase | anaphase A | anaphase B |
| ARS element | ascus | aster |
| cdc2 kinase | cell cycle | cell plate |
| *CEN* sequence | centromere | centromere |
| centrosome | chiasmata | chromatid |
| conjugation | contractile ring | cytokinesis |
| diakinesis | diploid | endospore |
| $G_0$ state | $G_1$ phase | $G_2$ phase |
| gamete | gene conversion | growth factor |
| haploid | homologous chromosomes | homologous recombination |

165

| | | |
|---|---|---|
| INCENPs | interphase | karyotype |
| kinetochore | kinetochore microtubule | lamin |
| leptotene | M phase | MAP kinases |
| maturation promoting factor | meiosis | metaphase |
| midbody | mitosis | mitotic spindle |
| motor protein | *oriC* | pachytene |
| PDGF | phragmoplast | prokaryotic endospore |
| prophase | protein-tyrosine kinase | *Ras* |
| recombination | replication origin | replicon |
| restriction point | S phase | START |
| synapsis | synaptonemal complex | Tel sequence |
| telomerase | telophase | temperature-sensitive mutants |
| tetrad | transduction | transformation |
| zygotene | | |

# KEY FIGURE

Label the following:
   active MPF
   cdc2 kinase
   cyclin
   inactive MPF
   interphase
   mitosis

When does mitosis occur?

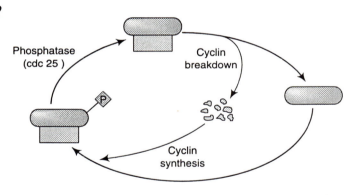

# Chapter 12

## 👉 STUDY QUESTIONS

1. If cells from one species of prokaryote are filtered through a membrane filter with a pore size of 2 μm, some cells will pass through the filter while others will remain on the filter. The purpose of this procedure is to
   a. Induce synchronous cell division.
   b. Select the cells that are ready to divide.
   c. Separate two species.
   d. Select temperature-sensitive mutants.
   e. None of the above.

2. *Mycobacterium tuberculosis*, the bacteria that cause tuberculosis, take 6 hr. for a cell cycle. If, the C period = 40 min. and the D period = 20 min., how do you account for the remaining time?
   a. *M. tuberculosis* has more than one C and D period.
   b. *M. tuberculosis* has a long D period.
   c. *M. tuberculosis* starts new DNA replication before completing the previous replication.
   d. *M. tuberculosis* has a B period.
   e. None of the above.

3. All of the following conditions are required for prokaryotic cell division **except**
   a. Availability of nutrients.
   b. ppGpp accumulates.
   c. DnaA is transcribed.
   d. DnaA unwinds DNA at *oriC*.
   e. None of the above.

4. Which one of the following occurs in the eukaryotic cell cycle and **not** in the prokaryotic cell cycle?
   a. Mitosis
   b. Chromosome replication
   c. Cell growth
   d. Cell division
   e. None of the above

5. If a cell in $G_1$ is fused with a cell in S,
   a. The S nucleus is arrested.
   b. The $G_1$ nucleus ~~starts chromosome replication.~~ enters the S phase
   c. The S nucleus enters $G_2$.
   d. Both nuclei start dividing.
   e. ~~None~~ Two of the above are correct

6. All of the following occur if MPF is injected into an $G_1$-phase cell **except**
   a. The nuclear envelope breaks down.
   b. The chromosomes condense.
   c. A mitotic spindle forms.
   d. The chromosomes are replicated.
   e. ~~None of the above.~~ Transcription ceases

167

7. All of the following occur to induce cell division **except**
   a. PDGF binds to a cell-surface receptor.
   b. Protein-tyrosine kinase is phosphorylated.
   c. *Ras* is activated.
   d. MAP kinases actiate AP1 transcription factor.
   e. None of the above.

8. Which one of the following is not correctly matched?
   a. ARS element—self-replicating DNA
   b. *CEN* sequence—attachment for two chromatids
   c. Tel sequence—Satellite DNA
   d. Topoisomerase II—replicating DNA
   e. None of the above

9. All of the following are involved in the M phase **except**
   a. Centrosome.
   b. H1.
   c. Kinetochore.
   d. Microtubule.
   e. MPF.

10. If a kinetochore on one side of a metaphase chromosome is destroyed during metaphase, the chromosome
    a. Moves to the opposite pole.
    b. Stops in place.
    c. Stops in the center of the cell.
    d. Attaches to another microtubule.
    e. None of the above.

11. All of the following occur in plant cytokinesis **except**
    a. The phragmoplast forms.
    b. The plasma membrane invaginates.
    c. ~~Membrane~~ vesicles accumulate.  [Golgi]
    d. The cell plate forms.
    e. A band of microtubules forms.

12. Structure A is
    a. A cleavage furrow.
    b. Peptidoglycan.
    c. The cell plate.
    d. Actin filaments.
    e. None of the above.

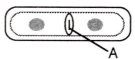

13. Structure B is
    a. A cleavage furrow.
    b. Peptidoglycan.
    c. The cell plate.
    d. Actin filaments.
    e. None of the above.

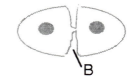

14. Which one of the following is **not** correctly matched?
    a. Conjugation—genetic recombination requiring cell-to-cell contact
    b. Transformation—genetic recombination via naked DNA
    c. Transduction—genetic recombination mediated by a virus
    d. Meiosis—genetic recombination between two opposite mating strains
    e. None of the above

15. The following stages occur in prophase I. What is the first stage?
    a. Diakinesis
    b. Diplotene
    c. Leptotene
    d. Pachytene
    e. Zygotene

16. A mutation that inactivates the *cdc25* gene will result in a cell that
    a. Divides too soon.
    b. Divides too many times.
    c. Elongates without dividing.
    d. Is in $G_0$ arrest.
    e. Has too many chromosomes.

17. If the cell shown on the left is in $G_1$, the cell on the right is in

    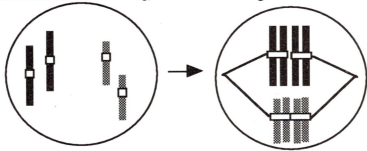

    a. $G_2$
    b. Metaphase I
    c. Metaphase II
    d. Metaphase of mitosis
    e. None of the above

18. The diploid number for bees is 16, male bees are haploid. During meiosis, how many tetrads form in a male's cells?
    a. 0
    b. 4
    c. 8
    d. 16
    e. 32

Use the following diagram to answer questions 19-21.

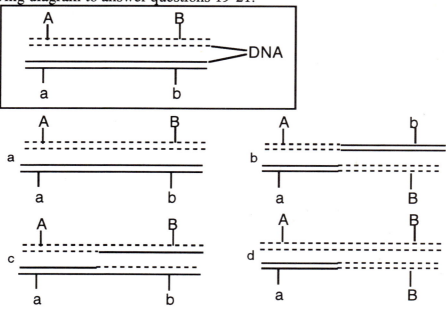

19. If reciprocal homologous recombination occurs between the two chromosomes shown at the top of the figure, which genotype should the resulting cell have?

20. If homologous recombination occurs between the two chromosomes shown at the top of the figure, which of the choices best depicts a Holliday junction?

21. Which choice best depicts gene conversion?

## PROBLEMS

1. *Clostridium* bacteria are anaerobes that lack catalase. Of what advantage are endospores to this bacterium? *C. botulinum* bacteria cause botulism. How do endospores contribute to their ability to cause botulism?

2. The synthetic auxin 2,4-dichlorophenoxyacetic acid (2,4-D) is widely used as a herbicide. Explain how a plant *growth* hormone can kill plants.

3. Taxol is an extract of the yew tree that is useful in treating some types of cancer. Taxol inhibits depolymerization of microtubules. How would this stop cancer?

4. The following experiment was done to determine whether the kinetochore duplicates in interphase I or interphase II. A pair of sister chromatids in meiosis II was transferred to a meiosis I spindle, the sister chromatids separated when meiosis continued. A tetrad in meiosis I was transferred to a meiosis II spindle, the tetrad separated into two homologous pairs when meiosis continued. From this information, when does the kinetochore duplicate?

5. The following cell cultures were exposed to $^3$H-thymidine and examined by autoradiography for labeled nuclei. What can you conclude from these data?

| Cell type | Labeled nuclei after exposure to $^3$H-thymidine, hr. | | | | | | | |
|---|---|---|---|---|---|---|---|---|
| | 11 | 12 | 13 | 14 | 15 | 16 | 17 | 18 |
| Human diploid fibroblasts | - | - | - | - | + | + | + | + |
| HeLa | - | - | - | - | - | - | - | + |
| Chinese hamster lung cells | + | + | + | + | + | + | + | + |

6. Identify where mitosis and meiosis occur during the yeast life cycle shown below. Haploid cells are shaded; diploid cells are white.

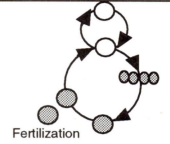

7. Cell A is a diploid cell formed from the conjugation of haploid cells. The genotypes of the haploid cells were: A+, B+, C+ × A-, B-, C-. Cell A went through meiosis to produce cells 1, 2, 3, and 4. These four cells then went through mitosis. Cell 1 produced cells 1 and 1′, cell 2 produced 2 and 2′, etc. The phenotypes of the cells are given in the table below.

| Cell | Phenotype. Observed | Expected with reciprocal recombination |
|---|---|---|
| 1 | ABC | ABC |
| 1′ | ABC | ABC |
| 4 | abc | abc |
| 4′ | abc | abc |
| 2 | AbC | Abc |
| 2′ | Abc | Abc |
| 3 | aBC | aBC |
| 3′ | aBc | aBC |

Are cells 1-4 and 1′-4′ haploid or diploid? What can you conclude from the phenotypes?

## ANSWERS TO STUDY QUESTIONS

1. b    2. d    3. b    4. a

5. b    6. d.   7. e    8. d

9. e    10. a   11. b   12. c

13. a   14. d   15. c   16. c

17. b   18. a   19. b   20. c

21. d

# CHAPTER 13

## *The Cytoskeleton and Cell Motility*

### 🔖 LEARNING OBJECTIVES

Be able to
1. Describe the arrangement of actin, myosin, troponin, tropomyosin, T tubules, and the sarcoplasmic reticulum in a sarcomere.
2. List the events of muscle contraction.
3. Identify the roles of ATP, $Ca^{2+}$, and phosphocreatine in muscle contraction.
4. Differentiate skeletal, smooth, and cardiac muscle.
5. Describe how a gel-sol transition can cause cell movement.
6. Describe the role of actin filaments in cell crawling, chemotaxis, and microvilli.
7. Describe the molecular structure of each of the following:
    a. Microtubules
    b. 9 + 2 flagella
    c. Centriole
    d. Centrosome
8. Describe the role of centrioles in a cell.
9. Describe the role of intermediate filaments in a cell.
10. List the functions of the cytoskeleton.
11. Compare and contrast the following pairs of terms:
    a. Eukaryotic and prokaryotic flagella
    b. Eukaryotic and prokaryotic basal bodies

### 🔖 CHAPTER OVERVIEW

**I. INTRODUCTION, pp. 563-564**
  1. The cytoskeleton provides support and motility.
  2. The cytoskeleton is built from actin filaments, microtubules, and intermediate filaments.

**II. MUSCLE CONTRACTION, pp. 564-582**

  **A. Skeletal Muscle Cells Contain Myofibrils Constructed from an Organized Array of Thick and Thin Filaments, pp. 564-565**
  1. A muscle fiber measures 10 to 100 µm in diameter and several centimeters in length.
  2. Muscle fibers arise from embryonic myoblasts which fuse to form giant cells.
  3. Muscle fibers are multinucleated and surrounded by a plasma membrane called the sarcolemma.
  4. Myofibrils, found in the cytoplasm, are the contractile structures.
  5. Z disks divide a myofibril into 2-µm sarcomeres.
  6. I bands contain thin filaments (~6 nm diameter) and A bands contain thick filaments (~15 nm diameter).
  7. Crossbridges protrude from thick filaments and connect to thin filaments.

  **B. Actin and Myosin Are the Main Constituents of Myofibrils, p. 565**

173

# Chapter 13

**C. Myosin II Is the Principal Protein of the Thick Filament, p. 566**
  1. Myosin consists of six polypeptide subunits, two heavy chains and four light chains.
  2. Myosin II is a long thin rod with two globular heads at one end found in myofibrils.
  3. In myosin II the two α-helical heavy chains form a coiled coil.
  4. Myosin I is involved in mytoplasmic movements in nonmuscle cells.

**D. Myosin Hydrolyzes ATP, Binds to Actin, and Polymerizes into Filaments, pp. 567-568**
  1. Myosin hydrolyzes ATP in the presence of $Ca^{2+}$.
  2. Each globular head is an ATPase and has an actin-binding site.
  3. Myosin:
     a. Hydrolyzes ATP
     b. Binds to actin
     c. Binds another ATP
     d. Dissociates from actin
  4. Myosin polymerizes into a bipolar structure with the globular heads extending in opposite directions.

**E. Actin Is the Principal Protein of the Thin Filament, p. 568**
  1. Actin filaments consist of α-actin molecules assembled at angles to look like a helix.
  2. Actin filaments exhibit polarity: the plus end grows fastest; the minus end points away from the Z disk.

**F. Muscle Contraction Is Caused by the Sliding of Thick and Thin Filaments, pp. 568-571**
  1. According to the sliding filament model, myosin crossbridges at each end of a myosin filament and pulls the actin thin filaments toward the center.

**G. Filament Sliding Is Powered by ATP-Driven Changes in the Myosin Head, pp. 571-572**
  1. Energy from ATP causes a conformational change in the myosin head so it can bind further along the actin filament.
  2. The steps in contraction are:
     a. Myosin hydrolyzes ATP.
     b. $P_i$ dissociates from myosin and myosin binds actin.
     c. Myosin causes the sliding of the attached actin filament.
     d. ADP is released and myosin binds another ATP.

**H. T Tubules Transmit Incoming Action Potentials to the Interior of the Skeletal Muscle Cell, pp. 572-574**
  1. Nerve axons meet muscles at the neuromuscular junction.
  2. Axon branches come near (50 nm) junctional folds in the sarcolemma.
  3. Acetylcholine released from a nerve binds with receptors in the junctional folds.
  4. The receptors function as gated channels permitting $Na^+$ into the muscle cell.
  5. If depolarization exceeds a threshold level, an action potential is triggered.
  6. In 1958, Huxley and Taylor applied electric current to selected regions on the surface of a frog muscle cell. They found especially sensitive spots located over I bands.

## Chapter 13

    7. Membrane channels called transverse tubules (T tubules) are located at the I bands; occasionally making contact with the sarcolemma with their lumens open to the outside.

**I. T Tubules Trigger Muscle Contraction by Promoting the Release of $Ca^{2+}$ from the Sarcoplasmic Reticulum, pp. 574-575**
1. The sarcoplasmic reticulum is a system of flattened membrane channels covering the surface of each myofibril and enlarged to form terminal cisternae near the Z bands.
2. The calcium pump of the sarcoplasmic reticulum takes up $Ca^{2+}$ using ATP.
3. $Ca^{2+}$ is trapped in the sarcoplasmic reticulum lumen, bound to calsequestrin.
4. When an action potential is transmitted by T tubules from the cell surface to the center of a triad, the membrane depolarization causes an increase in premeability to $Ca^{2+}$ which diffuses into the cytosol.

**J. Troponin and Tropomyosin Make Filament Sliding Sensitive to $Ca^{2+}$, pp. 575-576**
1. Tropomyosin is a dimer of α and β subunits.
2. Troponin consists of troponin-T, troponin-I, and troponin-C.
3. Tropomyosin binds to actin to block sites where myosin attaches—to keep the myofibril relaxed.
4. When the sacroplasmic reticulum releases $Ca^{2+}$, $Ca^{2+}$ binds to troponin and the troponin displaces the tropomyosin to permit actin-myosin binding.
5. When the depolarization signals are gone, the sarcoplasmic reticulum takes up $Ca^{2+}$, and tropomyosin binds actin.

**K. Phosphocreatine Replenishes ATP during Muscle Contraction, pp. 577-578**
1. ATP for the calcium pump and myosin is provided by the reaction:

Phosphocreatine + ADP $\xrightarrow{\text{creatine kinase}}$ ATP + creatine.

**L. Muscles That Frequently Contract Contain Large Numbers of Mitochondria, pp. 578-579**
1. Mitochondria can occupy as much as 50% of the total cytoplasmic volume in muscle cells.
2. Muscles designed for long, continuous use contain the oxygen-storing pigment, myoglobin.
3. Red muscle (or "dark meat") is colored by myoglobin and the mitochondrial cytochromes.
4. Red muscle stores fat, the mitochondria oxidize the fatty acids to produce ATP or phosphocreatine.
5. White muscle has fewer mitochondria and stores glycogen.
6. White muscle gets ATP from glycolysis.

**M. Several Proteins Help to Maintain the Structural Integrity and Metabolic Efficiency of the Myofibril, p. 579**
1. Titan forms an elastic connection between the Z disk and myosin to maintain structural integrity in relaxed muscles.
2. α-actinin crosslinks actin filaments.
3. Nebulin attaches actin filaments to the Z disk.
4. M protein links myosin at the M line.
5. D protein holds myosin filaments.
6. Desmin attaches Z disks of adjacent myofibrils.

| Proteins in muscle cells | Percentage of protein content of myofibril |
|---|---|
| Myosin | 50 |
| Actin | 25 |
| Troponin and tropomyosin | 5 |
| Other proteins | 20 |

- N. **Defects in Skeletal Muscle Cells Lead to Myasthenia Gravis and Muscular Dystrophy, pp. 579-580**
    1. Myasthenia gravis is an autoimmune disease that may be due to a decline in the number of acetylcholine receptors in the sarcolemma.
    2. Duchenne muscular dystrophy is inherited on the X chromosome. It is characterized by progressive degeneration of skeletal muscle cells.
    3. Duchenne muscular dystrophy gene codes for an abnormal actin-binding protein, dystrophin.

- O. **Cardiac Muscle Is Composed of Cells That Are Joined Together by Intercalated Disks, pp. 580-581**
    1. The cardiac muscle is divided into uninucleated cells by intercalated disks.
    2. Intercalated disks are specialized plasma membranes containing α-actinin and vinculin and gap junctions.
    3. Cardiac muscle has numerous mitochondria and a prolonged depolarization phase to allow all the cells to contract as one unit.

- P. **Smooth Muscle Cells Lack the Highly Organized Array of Filaments Seen in Skeletal and Cardiac Muscle, p. 581**
    1. Smooth muscle cells are uninucleated and lack the striations of skeletal muscle.
    2. The ratio of actin to myosin is 15 to 1.
    3. Thin and thick filaments are in parallel groups randomly dispersed throughout the cytoplasm.
    4. Thin filaments are linked by α-actinin to dense bodies and to the plasma membrane at focal adhesions.

- Q. **Myosin Light-Chain Kinase Controls Smooth Muscle Contraction, pp. 581-582**
    1. Smooth muscles lack troponin and tropomyosin.
    2. $Ca^{2+}$ binds to calmodulin which activates myosin light-chain kinase.
    3. The kinase phosphorylates the myosin light-chain to permit myosin-actin binding.
    4. Myosin light-chain phosphatase removes the phosphate from the myosin.
    5. Smooth muscle contraction is initiated by nerves or hormones.

III. **ACTIN FILAMENTS IN NONMUSCLE CELLS, pp. 582-595**
  1. Microfilaments in the cytoplasm of most eukaryotic cells are actin filaments; they make up 10-15% of the cell's total protein mass.

- A. **Actin-Binding Proteins Influence the Structural Organization of Actin Filaments, pp. 582-584**
    1. Nonmuscle cells have β-actin and γ-actin.
    2. Actin-binding proteins, controlled by $Ca^{2+}$, $Ca^{2+}$-calmodulin, or $PIP_2$ interact with actin.
        a. Monomer-binding proteins prevent actin filament formation.
        b. Bundling proteins create parallel bundles of actin.
        c. Gelating proteins form less regular crosslinks between actin filaments.

d. Capping and severing proteins prevent growth of filaments or cause filaments to break.
e. Motor proteins (e.g., myosin) cause movement.
f. Regulatory proteins (e.g., tropomyosin) influence actin's interactions.
g. Anchoring proteins link actin to membranes and other filaments.
3. Cytochalasin binds actin and inhibits filament formation.
4. Phalloidin promotes actin polymerization.

**B. The Actin Filaments of Nonmuscle Cells Are Often Anchored to the Plasma Membrane, p. 584**
1. Stress fibers are parallel bundles of actin filaments attached to the plasma membrane at focal adhesions.
2. Stress fibers may contain myosin and tropomyosin and may function in movement.
3. Actin filaments at the outermost region of the cell (the cell cortex) add mechanical strength and movement.

**C. Cytoplasmic Streaming Is Driven by Interactions between Actin Filaments and Myosin, pp. 584-586**
1. Brownian movement is generated by the thermal energy of water molecules.
2. Directed movement of the cytoplasm is cytoplasmic streaming.
3. In plant cells, cytoplasmic streaming follows a circular path or cyclosis.
4. Parallel rows of actin filaments lie at the interface between the outer immobile (cortical) cytoplasm and the inner, moving cytoplasm.
5. The plus ends of the actin are pointed in the direction of streaming, suggesting that the movement of myosin is involved.
6. Cytoplasmic streaming in animal cells is called shuttle streaming; it can cause shape changes and movement in the cells.

**D. Cell Crawling in Amoeba Depends on Cytoplasmic Streaming Driven by Forces Generated at the Front of the Cell, pp. 586-587**
1. Cell crawling (amoeboid motion) moves a cell along a solid surface.
2. Cells form projections called pseudopodia at the advancing edge.
3. Endoplasm (cytoplasm in the cell's interior) flows from the tail toward the pseudopodium.
4. At the pseudopodium, the endoplasm is diverted to the edge and becomes more rigid cytoplasm called ectoplasm.
5. Ectoplasm now liquefies and moves forward.
6. Ectoplasm is rigid because actin filaments form (gel state).
7. Fragmentation of actin (sol state) forms the more fluid endoplasm.

**E. Actin-Binding Proteins Mediate Sol-Gel Transitions, Gel-Sol Transitions, and the Contraction of Actin Gels, pp. 587-588**
1. Sol-gel transitions: Filamin or α-actinin forms crosslinks between actin and creates a gel.
2. Gel-sol transitions: Actin-binding proteins that break actin filaments will create a sol.
3. Contraction of actin gels occurs in the presence of $Ca^{2+}$ and myosin.

**F. Cell Crawling in Multicellular Animals Is Based on Repeated Cycles of Extension, Attachment, and Contraction, pp. 588-590**
1. Cell crawling involves:
a. Extension of the leading edge of the cell. Extensions called lamellipodia, microspikes, filopodia, or lobopodia form by growing actin filaments.

The filament grows by adding monomers to the plus end and taking them away from the minus end (treadmilling).
- b. Attachment of the leading edge to the substratum. Focal adhesions form between fibronectin in the extracellular matrix and fibronectin receptors in the cell membrane. When an extension binds tightly to the substratum, the cell moves in its direction.
- c. Contraction of the cytoplasm. Increased [$Ca^{2+}$] at the tip of a pseudopodium or actin-severing proteins causes a wave of contraction.

G. **Crawling Cells Are Guided by Chemical Attractants and Repellents, pp. 590-591**
1. The movement of a cell toward or away from a substance is chemotaxis.
2. An attractant binds a cell surface receptor and actin polymerization causes the formation of cell extensions on the surface facing the attractant.

H. **Microvilli and Stereocilia Are Supported by Bundles of Crosslinked Actin Filaments, pp. 591-592**
1. Microvilli contain a bundle of actin filaments for support; the plus ends point toward the tip of the microvillus.
2. The microvilli form because the actin filaments are crosslinked with villin.
3. The actin bundle adheres to the plasma membrane at the tip of the microvillus and to the terminal web at the minus end.
4. Stereocilia projecting from epithelial cells in the inner ear are sensitive to sound vibrations and fluid motion in the semicircular canals.
5. Movements alter the permeability of ion channels in the plasma membrane.

I. **Actin Filaments Influence the Mobility of Plasma Membrane Proteins, pp. 592-593**
1. Cell surface receptors can collect into a cap.
2. Actin filaments accumulate underneath the cap suggesting that the receptors are guided by the actin filaments.
3. In red blood cells, spectrin is crosslinked by short actin filaments just under the plasma membrane.
4. Membrane proteins move in accordance with movements induced in the actin-spectrin meshwork.

J. **Cell Cleavage by the Contractile Ring Is Based on an Interaction between Actin and Myosin, p. 593**
1. The contractile ring is composed of actin.
2. Myosin is necessary for cytokinesis because cells lacking myosin cannot divide.

K. **Actin Filaments Influence the Shape of Animal Cells, p. 594**
1. On contact with a solid surface, spherical cells produce actin filament-containing microvilli which elongate into filopodia then spread out to produce a thinly flattened cell.

L. **Phagocytosis and Exocytosis Require the Participation of Actin Filaments, pp. 594-595**

# Chapter 13

IV. **MICROTUBULES, pp. 595-619**

A. **Microtubules Are Long Hollow Cylinders Constructed from the Protein Tubulin, pp. 595-596**
 1. Microtubules are a hollow core surrounded by 13 protofilaments.
 2. Protofilaments are made of tubulin.
 3. Tubulin is a heterodimer consisting of α-tubulin and β-tubulin.
 4. The heterodimers are arranged in helix.

B. **Microtubule Assembly Is Promoted by Microtubule-Associated Proteins (MAPs), pp. 596-597**
 1. Addition of MAPs promotes tubule formation.
 2. Assembly of microtubules from ring forms is faster than from heterodimers.
 3. Phosphorylation of MAPs controls tubule formation.

C. **A Microtubule Is A Polar Structure That Grows More Rapidly at its Plus End, pp. 597-599**
 1. The growing end is called the plus end.
 2. Treadmilling can occur also.

D. **GTP Hydrolysis Contributes to Dynamic Instability of Microtubules, pp. 599-600**
 1. GTP can bind to α-tubulin and β-tubulin.
 2. GTP on the β-tubulin is hydrolyzed when a heterodimer is added to a growing microtubule.
 3. Microtubules exhibit dynamic instability:
    a. High GTP and unpolymerized tubulin → growing phase
    b. Low GTP → depolymerization.

E. **Centrosomes Serve as Initiation Sites for Microtubule Assembly in Animal Cells, pp. 600-602**
 1. Microtubules tend to form at MTOCs.
 2. In animal cells, the primary MTOC is the centrosome.
 3. Each growing microtubule is anchored to the centrosome at the minus end.
 4. Microtubule assembly may be initiated by γ-tubulin in the centrosome.

F. **Microtubule Stability Is Influenced by Various Enzymes, MAPs, and Drugs, pp. 602-603**
 1. In relatively static cells, microtubules are stabilized against depolymerization by STOP proteins at the plus ends.

G. **The Axoneme of Eukaryotic Cilia and Flagella Consists of a 9 + 2 Array of Linked Microtubules, pp. 603-606**
 1. Eukaryotic cilia and flagella are similar; they will be referred to as eukaryotic flagella.
 2. The axoneme is surrounded by the plasma membrane; it consists of 9 doublet tubules surrounding two central tubules; "9 pairs + 2."
 3. The 2 single tubules have 13 protofilaments.
 4. The doublet tubules are joined; the A-tubule has 13 protofilaments and the B-tubule has 10-11.
 5. The 9 pairs are connected by nexin links and the A-tubules are connected to the central region by radial spokes.
 6. Arms projecting from the outer doublets are dynein.

# Chapter 13

**H. The Motility of Eukaryotic Cilia and Flagella Is Based on Microtubule Sliding, pp. 606-609**

**I. Dynein Is a Motor Protein That Causes Microtubule Sliding, pp. 609-610**
1. Dynein is a motor protein.
2. Dynein has ATPase activity.
3. ATP hydrolysis causes the dynein arms on one A-tubule to bind and detach from sites on the adjacent B-tubule.
4. Changes in intracellular [$Ca^{2+}$] control the direction of swimming.
5. Flagella bend because dynein arms on one side of the axoneme are active while those on the other side are not.

**J. Cilia and Flagella Grow by Assembling Microtubules at the Axonemal Tip, pp. 610-611**
1. The plus end of the axoneme is at the top; in the doublets, new tubulin is added to the tip.
2. The two central tubules are capped where they attach to the plasma membrane and new tubulin is added to the base.

**K. Basal Bodies Initiate the Assembly of Cilia and Flagella, pp. 611-612**
1. The basal body is located at the base of every eukaryotic flagellum.
2. The basal body consists of nine sets of triplet microtubules: A-, B-, and C-tubules.
3. The A- and B-tubules are continuous with the A- and B-tubules in the flagellum.
4. The basal body may contain flagella genes.

**L. Basal Bodies Arise from Centrioles, pp. 612-613**
1. A centriole is composed of nine triplet microtubules.
2. Each centrosome contains two centrioles at right angles to each other.
3. A centriole migrates to the plasma membrane and initiates axoneme assembly; the centriole is then called a basal body.
4. In animal cells, centrioles are self-replicating.
5. In plant cells, centrioles arise from granular material that functions as a microtubule-organizing center.

**M. Ordered Microtubule Arrays Occur in Sensory Cilia and in Several Organelles Found in Unicellular Eukaryotes, pp. 613-614**
1. Sensory organs have nonmotile cilia; these cilia lack dynein.
2. Protozoan axostyles, cortical fibers, and cytopharyngeal baskets are composed of collections of single microtubules.

**N. Microtubules Are Involved in Moving Cytoplasmic Organelles and Other Intracellular Components, p. 614**
1. Microtubules move chromosomes during mitosis and meiosis.
2. Microtubules are responsible for saltatory movements (sudden, rapid movements) of organelles.

**O. The Involvement of Microtubules in Axonal Transport Can Be Observed Using Video-Enhanced Light Microscopy, pp. 614-616**
1. Slow axonal transport moves proteins and cytoskeletal filaments down the axon at 1-5 mm per day.

# Chapter 13

    2. Fast axonal transport moves vesicles and organelles in both directions in the axon at 100-500 mm per day.

**P. Kinesin and Dynein Propel Vesicles in Opposite Directions along Axonal Microtubules, pp. 616-617**
1. Kinesin is a motor protein used to transport vesicles from the cell body to the axon (minus-to-plus end).
2. Cytoplasmic dynein is responsible for transport in the opposite (retrograde) direction.

**Q. Cytoplasmic Microtubules Influence the Distribution of Cell Surface Components, pp. 617-619**
1. In plant cells, microtubules arrange the cellulose microfibrils for cell wall synthesis.

**R. Microtubules Influence Cell Shape and Internal Organization, p. 619**
1. In plant cells, placement of cellulose microfibrils by microtubules controls cell shape.
2. In animal cells, the position of microfibrils controls cell shape.
3. Position of the ER and Golgi complex is maintained by microtubules.

## V. INTERMEDIATE FILAMENTS, pp. 619-622

| Cytoskeletal element | Diameter |
| --- | --- |
| Actin | 6 nm |
| Intermediate filaments | 10 nm |
| Microtubules | 25 nm |

**A. Intermediate Filament Proteins Are Grouped into Six Main Classes, pp. 619-621**
1. Intermediate filaments (IF) are dimers consisting of two intertwined polypeptide chains.
2. IFs assemble from members of the same class.
3. Phosphorylation promotes the breakdown of filaments.

| Intermediate filament | Cell type |
| --- | --- |
| Acidic keratin | Epithelial cells |
| Neutral/basic keratin | Epithelial cells |
| Vimentin | Fibroblasts |
| Desmin | Muscle cells |
| Neurofilament proteins | Nerve cells |
| Lamins | Nuclear lamina |
| Nestin | CNS stem cells |

**B. Intermediate Filaments Provide Mechanical Strength and Support, pp. 621-622**
1. Keratin anchors the nuclear envelope and provides support at desmosomes.

## VI. CYTOSKELETAL INTERCONNECTIONS, pp. 622-623

**A. Axon Growth Requires Both Actin Filaments and Microtubules, p. 622**

# Chapter 13

B. **Eukaryotic Cytoplasm Is Supported and Organized by an Underlying Network of Interconnected Cytoskeletal Filaments, pp. 622-623**
1. Even "free" ribosomes are bound to the cytoskeleton when they are involved in protein synthesis.
2. The cell shape and membrane proteins remain in place after detergent treatment, indicating they are bound to the cytoskeleton.

VII. **BACTERIAL FLAGELLA, pp. 623-627**

A. **Bacterial Flagella Are Smaller and Less Complex Than Eukaryotic Flagella, pp. 623-624**
1. A bacterial flagellum consists of a filament attached to a hook and a basal body.
2. The basal body is a series of rings associated with the plasma membrane and, in gram-negative bacteria, the outer membrane.

B. **Bacterial Flagella Are Assembled by Adding Flagellin Monomers to the Flagellar Tip, pp. 624-625**
1. Bacterial flagella are made of flagellin monomers arranged in a helix.
2. Flagellin may be transported to the tip through the hollow central channel of the flagellum.

| Characteristic | Eukaryotic flagella | Prokaryotic flagella |
| --- | --- | --- |
| Diameter | 250 nm | 15 nm |
| Composition | 9 + 2 Microtubules | Flagellin polymer |
| Basal body | 9 microtubule triplets in cytoplasm | Rings associated with plasma membrane |
| Outer covering | Plasma membrane | No membrane |
| Energy source | ATP | Proton gradient |
| Movement | Microtubule sliding | Rotation of flagellum |

C. **Bacterial Flagella Are Rotated by a Motor That Is Powered by a Proton Gradient, pp. 625-626**
1. Bacterial flagella propel cells by acting as a rotating rod.
2. The proton gradient across the plasma membrane provides the energy—perhaps by acting on the rings embedded in the plasma membrane.

D. **The Direction of Bacterial Swimming Is Determined by the Balance between Clockwise and Counterclockwise Flagellar Rotation, pp. 626-627**
1. Random swimming includes tumbling produced by clockwise rotation.
2. Straight-line swimming is due to counterclockwise rotation.
3. With counterclockwise rotation, the multiple flagella are twisted into one bundle.
4. Attractants suppress clockwise rotation; repellents promote clockwise rotation (and tumbling).

E. **Chemotactic Signals Are Relayed by a Pathway That Involves Protein Phosphorylation, p. 627**
1. Four types of chemotaxis receptor proteins are located in the cell membrane of *E. coli*.
2. CheA kinase phosphorylates CheY which acts on the flagellar motor to promote clockwise rotation.

3. After binding an attractant, CheA kinase is inhibited so the bacteria swim in a straight line.
4. After binding an attractant, the receptor is methylated. Methylation depresses the ability to transmit signals to CheA so tumbling resumes.
5. The cell then adapts to this concentration of attractant and is now sensitive to a higher concentration so it can move toward the higher concentration.

# KEY TERMS

| | | |
|---|---|---|
| 9 + 2 | A band | actin |
| actin filament | actin-binding protein | AVEC microscopy |
| axoneme | axon | basal body |
| calcium pump | capping | cell cortex |
| cell crawling | centriole | centrosome |
| chemotaxis | cilia | coiled coil |
| cytochalasin | cytoplasmic streaming | cytoskeleton |
| dendrite | dense bodies | dynamic instability |
| dystrophin | ectoplasm | endoplasm |
| fast axonal transport | filopodia | fimbrin |
| flagella | flagellin | GTP cap |
| I band | intercalated disks | intermediate filaments |
| kinesin | lamellipodia | light-chain phosphatase |
| lobopodia | microspikes | microtubule |
| microtubule-associated protein | microtubule-organizing center | microvilli |
| minus end | motor protein | muscle fiber |
| muscular dystrophy | myasthenia gravis | myoglobin |
| myosin I | myosin II | myosin light-chain kinase |
| neuromuscular junction | phalloidin | phosphocreatine |
| plus end | pseudopodia | ruffling |
| saltatory movement | sarcolemma | sarcomere |
| sarcoplasmic reticulum | sliding filament model | slow axonal transport |
| spectrin | stereocilia | stress fiber |
| substratum | T tubules | taxol |
| terminal web | transverse tubules | treadmilling |
| triad | tropomyosin | troponin |
| tubulin | villin | Z disk |

# Chapter 13

## 👉 KEY FIGURE

Label the following: A, I, Z, and H bands, sacromere, actin, myosin. Show the appearance when the muscle is contracted.

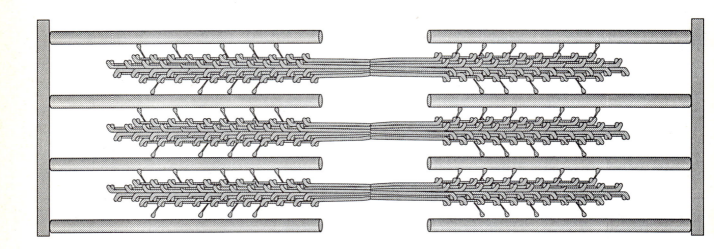

## 👉 STUDY QUESTIONS

1. Which letter indicates the A band?

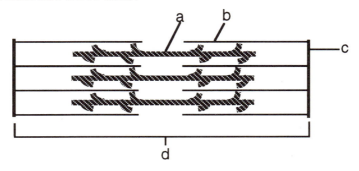

# Chapter 13

2. The viscosity of muscle extracts is decreased by the addition of ATP because ATP
    a. Is liquid.
    b. Causes actin and myosin to dissociate.
    c. Causes actin to dissolve.
    d. Causes myosin to dissolve.
    e. None of the above.

3. During muscle contraction
    a. A bands shorten.
    b. I bands shorten.
    c. Z ~~bands~~ discs shorten.
    d. ~~A bands lengthen.~~ M line disappears
    e. ~~I bands lengthen.~~ Thick filaments slide away from thin filaments

4. Contraction is not possible when a muscle fiber is stretched because
    a. Crossbridges can't reach the actin.
    b. Myosin collides with the Z band.
    c. ATP is not available.
    d. Myosin is bound to ADP.
    e. None of the above.

5. To which of the following does $Ca^{2+}$ bind to initiate muscle contraction?
    a. Tropomyosin
    b. Troponin
    c. Myosin
    d. Sacroplasmic reticulum
    e. Actin

6. An inhibitor of creatine kinase would
    a. Prevent muscle contraction.
    b. Result in a decrease in [ATP] in contracting muscles.
    c. Result in an increase in [phosphocreatine] in contracting muscles.
    d. Prevent $Ca^{2+}$ release.
    e. Prevent myosin crossbridges.

7. The Arctic tern has the longest migration of any bird, annually covering 22,000 miles. You would expect to find all of the following in the Arctic tern's breast muscles **except**
    a. Dark meat.
    b. Many mitochondria .
    c. Myoglogin.
    d. Stored glycogen.
    e. None of the above.

8. Which one of the following **leads** to all of the others?
    a. Activation of troponin
    b. Binding of mysoin to actin
    c. Depolarization of the sarcolemma
    d. Release of $Ca^{2+}$ from the sarcoplasmic reticulum
    e. Release of tropomyosin

# Chapter 13

9. The following events occur in smooth muscle contraction. What is the third step?
   a. $Ca^{2+}$ binds to calmodulin.
   b. Myosin light-chain kinase is activated.
   c. Myosin is phosphorylated.
   d. Myosin light-chain phosphatase acts on myosin.
   e. Intracellular $Ca^{2+}$ levels increase.

10. Inhibition of acetylcholine would stop contraction in all of the following **except**
    a. Skeletal muscle.
    b. Striated muscle.
    c. Smooth muscle.
    d. Cardiac muscle.
    e. None of the above.

11. During contraction, which band gets longer?

    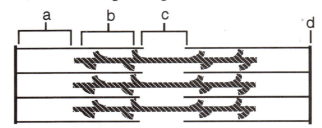

12. The poison phalloidin causes actin polmerization. This would cause cytoplasm to
    a. Gel.
    b. Sol.
    c. Contract.
    d. Crawl.
    e. None of the above.

13. The following events occur in cell crawling. What is the first step?
    a. Actin filaments are broken down.
    b. Actin filaments grow from the plus end.
    c. Fibrinonectin receptors bind the extracellular matrix.
    d. Lamellipodia form.
    e. The cytoplasm contracts.

14. All of following are required for microtubule assembly **except**
    a. GTP.
    b. Actin.
    c. Tubulin heterodimers.
    d. Phosphorylated MAPs.
    e. Centriole.

15. In the figure to the right, structure A is
    a. An axoneme.
    b. An actin filament.
    c. A pair of microtubules.
    d. Dynein.
    e. A centriole.

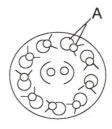

16. In the figure below, which letter is ATPase?

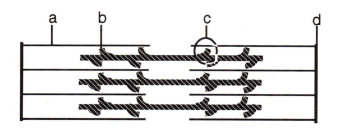

17. The most abundant protein in a myofibril is
    a. Actin.
    b. Dynein.
    c. Myosin.
    d. Troponin.
    e. Tropomyosin.

18. A human male who could not make dynein would be expected to
    a. Be sterile.
    b. Have recurrent respiratory infections.
    c. Have gastrointestinal problems.
    d. Have hearing problems.
    e. a and b

19. All of the following are characteristics of eukaryotic flagella **except**
    a. Tubules slide to extend flagellum.
    b. Flagellum is enclosed in the plasma membrane.
    c. Flagella rotate counterclockwise.
    d. Flagella are powered by ATP
    e. None of the above.

20. If a cell is treated with detergent which removes the plasma membrane, you would observe that
    a. Cell shape and membrane proteins remain in place.
    b. The cell's shape is greatly distorted and membrane proteins are in solution.
    c. The cell's shape is intact and membrane proteins are in solution.
    d. Organelles are in solution.
    e. None of the above.

21. All of the following are true about centrioles **except** they are
    a. Composed of 9 + 2 microtubules.
    b. Self-replicating.
    c. Continuous with tubules of flagella.
    d. Composed of tubulin.
    e. None of the above.

## PROBLEMS

1. When an animal dies, its body is flexible at first and then becomes increasingly rigid (*rigor mortis*). Following post-mortem rigidity, the body is again flexible as a taxidermist or mortician will attest. Explain what causes the temporary post-mortem rigidity and subsequent relaxation.

2. If the amoeba *Reticulomyxa* is treated with detergent to extract its organelles, the organelles will move (20μm/sec) when ATP is added. Explain how these organelles can move.

3. Whales can actively swim underwater for 30 minutes or more. Their lungs are not proportionally larger than human lungs. What muscle-cell adaptations do you expect to find in whales?

4. Some bacteria such as *E. coli* have peritrichous flagella, that is, the flagella are located over the entire surface of the bacterium. How can this bacterium swim in any direction? Other bacteria, such as *Vibrio* have a single polar flagellum. How can this bacterium tumble? What is the advantage of tumbling or random movement?

5. *Salmonella* bacteria adhere to receptors on epithelial cells of the small intestine. The bacteria enter the host cell and migrate through to the other side in an actin basket to be released into subepithelial tissue. Describe the molecular events required to move *Salmonella* bacteria through the epithelial cell.

6. Leukocyte adherence deficiency (LAD) is an inherited disease. People with LAD suffer from recurrent infections and fail to make pus although they have large numbers of white blood cells in their blood. Their white blood cells are lacking integrins. Explain how this deficiency leads to ineffective white blood cells.

7. Explain why you can turn a plant cell and gravity doesn't cause the organelles to fall to the bottom of the cell.

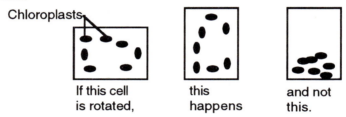

There is one exception to this; in root cells turned on their sides, statoliths fall to the new "bottom" of the cell. Statoliths are starch-containing vesicles. Explain why statoliths can fall while other organelles do not.

## ANSWERS TO STUDY QUESTIONS

| | | | | | | | |
|---|---|---|---|---|---|---|---|
| 1. | a | 2. | b | 3. | b. | 4. | a |
| 5. | b | 6. | b | 7. | d | 8. | c |
| 9. | b | 10. | c | 11. | b | 12. | a |
| 13. | b | 14. | b | 15. | c | 16. | b |
| 17. | c | 18. | e | 19. | c | 20. | a |
| 21. | a | | | | | | |

# CHAPTER 14

# *Evolution of Cells and Genetics of Cell Organelles*

## ⇗ LEARNING OBJECTIVES

Be able to
1. Compare and contrast the following:
    a. Cell theory and evolution theory
    b. Spontaneous generation and cell theory
2. Describe the contributions to the evolution theory of Oparin, Miller, S. Fox, de Duve, Woese, and Margulis.
3. Provide one explanation of how each of the following arose:
    a. Organic acids
    b. Proteins
    c. RNA
4. Define the following:
    a. Proteinoid
    b. Coacervate
    c. Microsphere
5. List the similarities and differences of the following:
    a. Eukaryotes
    b. Eubacteria
    c. Archaebacteria
6. Define maternal inheritance and biparental inheritance.
7. Compare and contrast the following pairs of terms:
    a. Mitochondrial DNA and chloroplast DNA
    b. Mitochondrial DNA and nuclear DNA
    c. Chloroplast DNA and nuclear DNA
8. List the organelles that develop from proplastids.
9. State the endosymbiotic theory.
10. Use the endosymbiotic theory to explain the origin of mitochondria and chloroplasts.

## ⇗ CHAPTER OVERVIEW

I. INTRODUCTION, p. 632
    1. Living ancient organisms are not available.
    2. Evolution experiments are designed
        a. To determine the kinds of events that would have been possible
        b. To explain features of modern cells.

II. THE FIRST CELLS, pp. 632-640

    A. **Biologists Once Believed That Cells Routinely Arise by Spontaneous Generation, pp. 632-633**
        1. Spontaneous generation is the idea that cells arise from nonliving matter.
        2. In 1861, Pasteur proved that (bacterial) cells arise from pre-existing cells.
        3. In 1855, Virchow stated that "all cells come from cells."
        4. Evolution theory says that the initial cells did arise spontaneously.

5. Macromolecules can assemble spontaneously and self-assembly has been observed in ribosomes, microtubules, actin filaments, membranes, and viruses.

B. **Fossil Evidence Suggests That Cells First Appeared on Earth About 3.5 Billion Years Ago, p. 633**
 1. The Earth is ~4.5 billion years old.
 2. The oldest fossils are ~3.5 billion years old.
 3. One hypothesis states that the first cells were brought to earth from a distant planet.

C. **Organic Molecules Form Spontaneously under Simulated Primitive Earth Conditions, pp. 633-635**
 1. In the 1920s, Oparin suggested that the primitive Earth was favorable for the spontaneous formation of organic molecules.
 2. In 1953, Miller simulated conditions of the early Earth:
    a. $CH_4$, $NH_3$, $H_2$, $H_2O$
    b. An electric spark was substituted for lightning.
    c. A variety of organic molecules including amino acids and fatty acids formed.
 3. The energy requirements for the condensation reactions necessary to form polymers could be met by:
    a. Anhydrous conditions and moderate heating.
    b. A condensing agent such as polyphosphate.
 4. De Duve proposed that thioesters acted as condensing agents.

D. **Catalysis and Information Transfer Were Essential for Development of the First Cells, pp. 635-636**
 1. How did the first catalysts arise?
    a. Proteinoids are amino acid polymers that form spontaneously in the absence of water.
    b. Proteinoids can catalyze hydrolysis, decarboxylation, redox, and amination reactions.
 2. How did the property of information transfer arise?
    a. RNA can form from A, U, C, and G without enzyme catalysis.

E. **Organic Molecules Spontaneously Generate Boundaries That Separate Them from the Surrounding Environment, pp. 636-638**
 1. Macromolecules in solution form coacervate droplets.
 2. Coacervates made of phosphorylase take up glucose phosphate, convert it into starch, and grow larger and divide.
 3. Coacervates require a high concentration of a particular macromolecule to form, they leak, and they are not stable.
 4. Microspheres are vesicles produced from proteinoids in the laboratory.
 5. Microspheres have the catalytic activity of proteinoids and can grow and divide.

F. **Did Proteins Arise Before Nucleic Acids? p. 638**
 1. Microspheres containing proteinoids would have had an advantage over unbound organic molecules.
 2. Proteinoids can bind to RNA.

G. **Did Nucleic Acids Arise Before Proteins? pp. 638-639**
 1. RNA can form spontaneously, however, it cannot replicate itself.

2. RNA has catalytic activity.
3. RNA could direct amino-acid polymerization as amino acids bind to specific base sequences.

### H. Ancient Cells Branched into Eubacteria, Archaebacteria, and Eukaryotes Early in Evolution, pp. 639-640
1. The oldest fossils resemble prokaryotic cells.
2. Closely related organisms will have similar nucleotide sequences in their DNA and RNA; distantly related organisms will have few similarities.
3. Living cells can be divided into three groups based on rRNA sequences:
   a. Eubacteria. The oldest group, arising 500 million years before other cell types.
   b. Eukaryotes
   c. Archaebacteria which live in harsh environments. They are divided into methanobacteria, halobacteria, and sulfobacteria.
4. The rRNAs of these three groups have the same stem-and-loop sequences suggesting evolution from a common ancestor.

### I. How Did the Internal Membranes and Organelles of Eukaryotic Cells Arise? p. 640
1. The ancestral eukaryote, called urkaryote, evolved into a protoeukaryote.
2. A protoeukaryote has an enveloped nucleus and internal membranes; it lacks mitochondria and chloroplasts.

## III. GENETIC SYSTEMS OF MITOCHONDRIA AND CHLOROPLASTS, pp. 640-663

### A. Mutations Affecting Chloroplast Pigmentation Led to the Discovery of Chloroplast Genes, pp. 640-642
1. In 1909, crosses between normal and colorless *Mirabilis jalapa* revealed maternal inheritance, i.e., genes are contributed by the female parent.
2. Crosses between normal and colorless *Pelargonium* revealed biparental inheritance, i.e., genes for a trait are contributed by one parent.
3. These examples demonstrate cytoplasmic inheritance involving chloroplast pigmentation.
4. The *iojap* gene in maize is a nuclear gene that causes a mutation in the cell's chloroplasts.

### B. The *Petite* and *Poky* Mutations Led to the Discovery of Mitochondrial Genes, pp. 642-643
1. *Petite* yeast lack cytochromes. They grow slowly and produce small colonies because they only use fermentation.
2. *Neutral petite* yeast × normal yeast → normal offspring.
3. *Suppressive petite* yeast × normal yeast → *petite* offspring.
4. *Poky Neurospora* have abnormal mitochondria and lack cytochromes *b, a,* and $a_3$.
5. *Poky* males × normal females → normal offspring.
6. *Poky* females × normal males → *poky* offspring.

### C. Mitochondria and Chloroplasts Contain Their Own DNA, pp. 643-645
1. DNA in organelles was first discovered in the kinetoplast (mitochondria) of trypanosomes.

# Chapter 14

2. Isodensity centrifugation is used to separate mitochondrial, chloroplast, and nuclear DNA.

**D. Why Is Cytoplasmic Inheritance Usually Maternal? p. 646**
1. Organelles are typically from the female gamete in organisms where the male gamete contains little cytoplasm.
2. In *Chlamydomonas*, male and female gametes are the same size and female chloroplast DNA is inherited.
3. DNA in the organelles of the female gamete is marked by methylation prior to fertilization; DNA from the paternal chloroplast is destroyed in the zygote.

**E. Mitochondrial DNA Is Usually Circular, pp. 646-648**
1. The amount of mitochondrial DNA varies:

|  | Percentage of total DNA contributed by mitochondria |
|---|---|
| Typical range | 0.1-1.0% |
| Yeast | 15 |
| Amphibian oocytes | 90 |

2. Mitochondrial DNA lacks histones and is generally circular.
3. Most mitochondria have more than one DNA molecule.

|  | Number of DNA molecules/mitochondrion |
|---|---|
| Animal | <Several hundred identical molecules |
| Fungi | <Several hundred identical molecules |
| Protists | <Several hundred identical molecules |
| Plants | 1 large master chromosome and two smaller molecules containing some of the master's sequences |

**F. Chloroplast DNA Is Also Circular and Present in Multiple Copies, pp. 648-649**
1. Chloroplasts contain 100-200 identical DNA molecules.

**G. Mitochondrial and Chloroplast DNA Replicate by a Semiconservative Mechanism Involving the Formation of D Loops, pp. 649-650**
1. Mitochondrial and chloroplast DNA are synthesized independent of S phase.
2. DNA replication is initiated at different locations on the complementary strands of DNA (called heavy and light strands).
3. The heavy strand is replicated faster, and the new strands form a D (displacement) loop.
4. In chloroplasts, DNA replication is initiated at different sites and the heavy and light strands are replicated at the same rate to produce two D loops.

**H. Mitochondrial DNA Codes for Ribosomal RNAs, Transfer RNAs, and a Small Group of Mitochondrial Polypeptides, pp. 650-652**
1. Human mitochondrial DNA contains 37 genes:

| Two coding for 16S and 22S rRNA | Heavy strand |
|---|---|
| 22 coding for tRNAs | 8 on light strand |
| 13 coding for inner mitochondrial membrane polypeptides | 12 on heavy strand |
| Replication origin | Heavy strand |

2. The polypeptide-genes code for NADH dehydrogenase, cytochrome *b*, cytochrome oxidase subunits, and ATP synthase subunits.
3. Mitochondrial DNAs of protists, fungi, and plants are larger than that of animal cells and includes noncoding sequences between genes and introns.

# Chapter 14

**I. Chloroplast DNA Codes for Over a Hundred RNAs and Polypeptides Involved in Chloroplast Gene Expression and Photosynthesis, p. 652**
1. Chloroplast DNA contains ≥ 120 genes coding for
   - rRNAs, tRNAs, ribosomal proteins, RNA polymerase
   - Thylakoid membrane polypeptides and a rubisco subunit
   - Unassigned reading frames (coding for unknown peptides)

**J. Both Strands of Human Mitochondrial DNA Are Completely Transcribed into Single RNA Molecules, pp. 652-654**
1. Transcription is initiated at a promoter in the D loop.
2. Both strands are completely transcribed and later cut into rRNAs, tRNAs, and mRNAs.
3. Mitochondrial RNase P cuts RNA at the 5′ end of tRNA to separate the RNAs.
4. Noncoding regions of the light-strand transcript are destroyed.

**K. Mitochondria and Chloroplasts Synthesize Proteins Using a Unique Set of Ribosomes, Transfer RNAs, and Protein Synthesis Factors, pp. 654-657**
1. Ribosomes:

| | | |
|---|---|---|
| Cytoplasmic | Eukaryotic | 80S |
|  | Prokaryotic | 70S |
| Mitochondria | Mammalian | 55-60S |
|  | Yeast | 75S |
|  | Higher plants | 78S |
| Chloroplasts |  | 70S |

2. tRNAs and aminoacyl-tRNA synthetases are different from their cytoplasmic counterparts.
3. Mitochondria and chloroplasts use formylmethionine.
4. The protein synthesis factors of mitochondria more closely resemble those in prokaryotes than in eukaryotic cytosol.
5. Inhibitors of protein synthesis:

|  | Inhibited by | | |
|---|---|---|---|
|  | Puromycin | Chloramphenicol | Cycloheximide |
| Prokaryotic | + | + | - |
| Eukaryotic | | | |
|   Cytoplasmic | + | - | + |
|   Mitochondrial | + | + | - |
|   Chloroplast | + | + | - |

**L. Mitochondria Employ a Slightly Altered Genetic Code, p. 657**

|  | Normal usage | Mitochondrial usage | |
|---|---|---|---|
|  |  | Mammals | Yeast |
| AUA | Ile | Met | Met |
| UGA | Stop | Trp | Trp |
| AGA, AGG | Arg | Stop | Arg |
| CUU, CUC, CUA, CUG | Leu | Leu | Thr |

**M. Mitochondria and Chloroplasts Synthesize Polypeptides Whose Genes Reside in Mitochondrial and Chloroplast DNA, pp. 657-658**

N. **Nuclear Genes Cooperate with Mitochondrial and Chloroplast Genes in Making Mitochondrial and Chloroplast Proteins, pp. 658-659**
   1. Mitochondria and chloroplasts have <10% of the genetic information needed to assemble either organelle.
   2. Most polypeptides are synthesized on free ribosomes and targeted for the organelles.
   3. The polypeptides contain signal sequences for receptors on the organelle's outer membrane.
   4. Some organelle proteins consist of subunits made in the cytoplasm and other subunits made in the organelle.
   5. Polypeptides made in the organelle are hydrophobic.

O. **New Mitochondria Arise by Growth and Division of Existing Mitochondria and Promitochondria, pp. 659-660**
   1. In most cells, mitochondria arise from growth and division of existing mitochondria.
   2. Yeast growing anaerobically have small double-membrane vesicles containing mitochondria DNA; they are called promitochondria.
   3. When exposed to oxygen, the promitochondria acquire cristae and cytochromes.

P. **New Chloroplasts Arise by Growth and Division of Existing Chloroplasts or Proplastids, pp. 661-663**
   1. Rapidly growing, undifferentiated plant tissue is called meristem.
   2. Meristem cells have proplastids that contain chloroplast DNA.
   3. Depending on location in the plant and light, proplastids develop into different kinds of plastids.

   | Plastid | Contains | Notes |
   |---|---|---|
   | Chloroplast | Chlorophyll, thylakoids | Develops in presence of light |
   | Etioplast | Protochlorophyllide | Develops in dark |
   | Chromoplast | Red, orange, yellow pigments | Flowers and fruits |
   | Proteinoplasts* | Proteins | |
   | Elaioplasts* | Lipids | |
   | Amyloplasts* | Starch | |
   | *Leucoplasts | Colorless | |

   4. In plant seedlings grown in the dark, proplastids develop prolamellar bodies and are called etioplasts.
   5. In plants transferred to the dark, thylakoid membranes breakup and form prolamellar bodies, and become etioplasts.
   6. Protochlorophyllide is a metabolic precursor of chlorophyll.
   $\xrightarrow{\text{5-aminolevulinate synthetase}}$ 5-aminolevulinate → Protochlorophyllide
   7. ALA synthetase is made after a brief exposure to red light and inhibited by exposure to far-red.
   8. The red/far-red response is mediated by phytochrome:
   Phytochome (inactive) $\underset{\text{Far-red } \lambda}{\overset{\text{Red } \lambda}{\rightleftarrows}}$ Phytochrome (active)
   9. Phytochrome probably stimulates transcription of certain genes.

# Chapter 14

**IV. ENDOSYMBIOSIS AND THE ORIGIN OF EUKARYOTIC ORGANELLES, pp. 663-669**

   **A. Mitochondria and Chloroplasts Exhibit Similarities to Bacterial Cells, pp. 664-665**
      1. DNA is circular and lacks histones
      2. Smaller ribosomes
      3. rRNA sequences
      4. fMet initiates protein synthesis
      5. Affected by antibiotics that inhibit protein synthesis
      6. Similar protein synthesis initiation and elongation factors
      7. Translate certain viral mRNAs
      8. RNA polymerases inhibited by rifampicin
      9. Similar superoxide dismutase

   **B. The Endosymbiotic Theory States That Mitochondria and Chloroplasts Evolved from Ancient Bacteria, pp. 665-667**
      1. According to the endosymbiotic theory, mitochondria and chloroplasts evolved from prokaryotic cells living symbiotically inside a protoeukaryote.
      2. Mitochondria appear to have evolved from an ancient purple eubacterium.
      3. Chloroplasts appear to have evolved from an ancient cyanobacterial type of eubacterium.

   **C. Did Peroxisomes, Cilia, and Flagella Also Arise by Endosymbosis? pp. 667-669**
      1. Peroxisomes lack DNA but they always arise from growth and division of existing peroxisomes.
      2. Cilia and flagella may have evolved from ancient spirochete bacteria.

   **D. The Cytomembrane System of Eukaryotic Cells May Have Arisen from Infoldings of the Plasma Membrane, p. 669**
      1. The ER, Golgi complex, lysosomes, secretion vesicles, endosomes, storage and transport vesicles, and nuclear envelope comprise the cytomembrane system because they are in communication with each other and to the plasma membrane.
      2. If an ancestral eubacterium lost the ability to synthesize muramic acid for a cell wall it could
         a. Develop an alternate cell wall—archaebacteria
         b. Develop an internal cytoskeleton for support—urkaryotes
      3. In wall-less urkaryotes, endocytosis could result in infoldings of the plasma membrane.

# ☞ KEY TERMS

| | | |
|---|---|---|
| amyloplast | archaebacteria | biparental inheritance |
| chloramphenicol | chromoplast | coacervation |
| condensation reaction | cytomembrane system | D loop |
| elaioplast | endosymbiotic theory | etioplast |
| eubacteria | eukaryotes | kinetoplast |
| maternal inheritance | maturase | meristem |
| microsphere | petite mutation | phytochrome |
| plastid | prolamellar body | promitochondria |
| proplastid | proteinoid | proteinoplast |
| protoeukaryote | RNA editing | self-assembly |
| urkaryote | | |

Chapter 14

# 🔑 KEY FIGURE

Label the archaebacteria, urkaryote, eubacteria, protoeukaryote, mitochondrion, and chloroplast. Show the events leading to the formation of the nucleus, mitochondrion, and chloroplast.

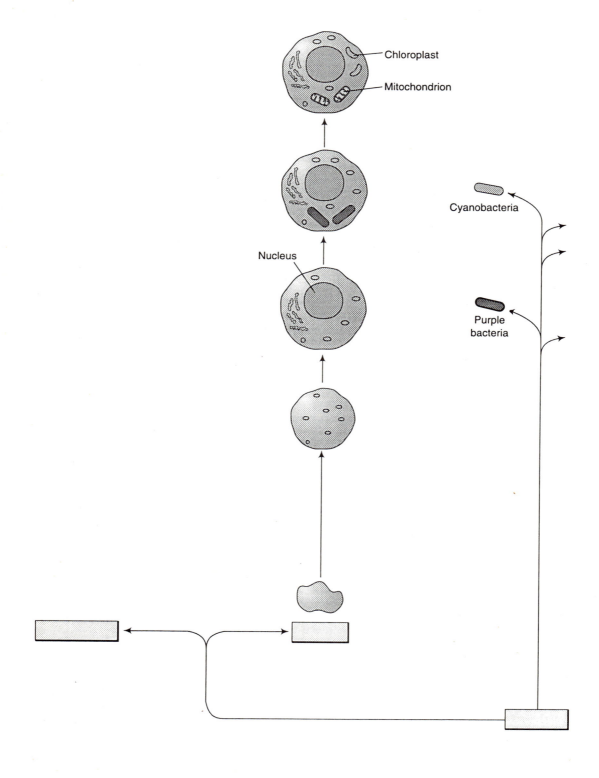

196

# Chapter 14

## STUDY QUESTIONS

1. The purpose of experiments looking at evolution is to
   a. Prove how life arose.
   b. Determine the types of events that were possible on the primitive Earth.
   c. Prove spontaneous generation.
   d. Disprove the idea of creation.
   e. All of the above.

2. The major distinction between spontaneous generation and evolution is that spontaneous generation addresses the origin of
   a. All life.
   b. Species.
   c. Individuals.
   d. Organelles.
   e. None of the above.

3. Which of the following was **not** part of Miller's experiment?
   a. Glucose
   b. $NH_3$
   c. $H_2$
   d. $CH_4$
   e. None of the above

4. The first genetic molecule was probably
   a. RNA.
   b. DNA.
   c. ATP.
   d. Protein.
   e. All of the above.

5. Which one of the following statements is **not** true?
   a. Proteinoids have catalytic activity.
   b. Coacervates can grow and divide.
   c. Microspheres have catalytic activity and can grow and divide.
   d. RNA is self-replicating and has catalytic activity.
   e. None of the above.

6. The first cells were probably
   a. Eukaryotes.
   b. Archaebacteria.
   c. Eubacteria.
   d. Proteinoids.
   e. None of the above.

7. The three cell types are differentiated on the basis of nucleotide sequences in their
   a. rRNA.
   b. tRNA.
   c. DNA.
   d. mRNA.
   e. ATP.

# Chapter 14

8. The inheritance of mitochondria is an example of
   a. Normal Mendelian inheritance.
   b. Maternal inheritance.
   c. Genetic recombination.
   d. Mutations.
   e. None of the above.

9. All of the following are true about mitochondrial DNA **except**
   a. Lacks histones.
   b. Is circular.
   c. Is packaged into nucleosomes.
   d. Is present in multiple copies.
   e. None of the above.

10. Cytoplasmic inheritance is usually maternal for all of the following reasons **except**
    a. Only females have chloroplasts.
    b. Female DNA is methylated.
    c. Nonmethylated DNA is destroyed in the zygote.
    d. Male gametes don't contain much cytoplasm.
    e. None of the above.

11. Which one of the following is **not** true?
    a. Chloroplast DNA is circular.
    b. Mitochondrial DNA is not arranged in nucleosomes.
    c. More than one copy of chloroplast DNA is present in a cell.
    d. Mitochondrial DNA is replicated in the nucleus.
    e. None of the above.

12. Proteins synthesized in the presence of cycloheximide are being made by
    a. Eukaryotic free ribosomes.
    b. Prokaryotic ribosomes.
    c. Mitochondrial ribosomes.
    d. b and c.
    e. Can't tell.

13. Mitochondrial DNA codes for all of the following **except**
    a. Mitochondrial RNA polymerase.
    b. Mitochondrial rRNA.
    c. Inner membrane proteins.
    d. Mitochondrial tRNA.
    e. None of the above.

14. Mitochondrial DNA can be altered by all of the following **except**
    a. RNA editing.
    b. Maturase.
    c. Nuclear mutations.
    d. Intron removal.
    e. None of the above.

15. Which one of the following statements is **not** true?
    a. Chloroplasts appear to have evolved from an ancient cyanobacterial type of eubacterium.
    b. Mitochondria appear to have evolved from an ancient purple eubacterium.
    c. Proplastids contain DNA, mRNA, rRNA, and tRNA.
    d. Promitochondria contain cytochromes.
    e. None of the above

16. Proplastids can develop into all of the following **except**
    a. Mitochondria.
    b. Chromoplasts.
    c. Proteinoplasts.
    d. Chloroplasts.
    e. Amyloplasts.

17. Mitochondria differ from their cell in all of the following **except**
    a. Genetic code.
    b. Inhibition by chloramphenicol.
    c. Start codon.
    d. Inhibition by puromycin.
    e. None of the above.

18. All of the following support the hypothesis for the endosymbiotic origin of peroxisomes **except**
    a. Peroxisomes can oxidize D-amino acids.
    b. Urate oxidase is absent in many organisms.
    c. Humans can survive without peroxisomes.
    d. The amino acid sequence of peroxisomal thiolase is similar to bacterial thiolase.
    e. None of the above.

19. Assume you have a *Euglena* culture that has been kept in the dark long enough to become colorless. Under which of the following conditions will the *Euglena* turn green?

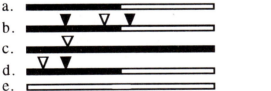 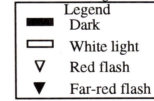

20. All of the following is evidence that centrioles arose from endosymbionts **except**
    a. Centrioles are self-replicating.
    b. Centrioles contain DNA.
    c. Centrioles contain microtubules.
    d. Bacteria have microtubules.
    e. Bacteria have DNA.

21. When frozen yeast cells are inoculated into a growth medium, they grow by fermentation because mitochondria need to be produced. Mitochondria are produced from
    a. Proplastids.
    b. Promitochondria.
    c. Etioplasts.
    d. Components in the cytoplasm.
    e. None of the above.

# Chapter 14

## 👉 PROBLEMS

1. Provide a model for formation of a protoeukaryote. (*Hint:* Read about the photosynthetic vesicles of purple and green bacteria in Chapter 9.)

2. To date the only evidence supporting the hypothesis that eukaryotic flagella may have evolved from symbiotic bacteria is observation of the protozoan *Mixotricha*. This protozoan is propelled entirely by bacteria living in specialized grooves on the protozoan's surface. What evidence is needed to support this hypothesis?

3. Colorless *Euglena* can ingest organic molecules and grow heterotrophically. *Euglena* loses its chloroplasts when it is (a) grown in the dark or (b) treated with streptomycin. When returned to the light, the (a) *Euglena* culture will produce chloroplasts but the (b) culture will never produce chloroplasts. Explain these results.

4. Some herbivorous sea slugs save the chloroplasts from their food in their digestive tracts. In the presence of light, the chloroplasts photosynthesize and produce sugars. Why do these animals still have to eat more algae?

5. Which type of metabolism, autotrophic or heterotrophic, most likely evolved first? Provide a rationale for your choice.

6. K. W. Jeon observed that a previously healthy amoeba culture became sick. On examination, 150,000 bacteria were inside each sick amoeba. Many amoeba died, however, some survived and grew again as healthy cells. Examination of these survivors showed 50,000 bacteria inside each amoeba. After 5 years, Jeon transplanted the nucleus from one of these amoeba to an enucleated, uninfected amoeba. The resulting cell could not divide. When the nucleus was put into an enucleated, infected amoeba, the resulting cell did divide. Explain.

7. Mutant mitochondria can be created in the laboratory. Commonly used mutations are $mit^-$, which is deficient in some respiratory enzymes, such as cytochrome oxidase, and $Amg^R$, which is resistant to the antibiotic aminoglycoside. ($Amg^S$ = sensitive to aminoglycoside.) Assume that all the mitochondria in a yeast cell are destroyed by streptomycin treatment and two new mitochondria, a $mit^--Amg^S$ and a $mit^+-Amg^R$, are put into the yeast cell. After the yeast grows, its progency have mitochondria that are $mit^+$ and $Amg^S$. Explain how these mitochondria occurred.

## 👉 ANSWERS TO STUDY QUESTIONS

| | | | | | | | |
|---|---|---|---|---|---|---|---|
| 1. | b | 2. | c | 3. | a | 4. | a |
| 5. | e | 6. | c | 7. | a | 8. | b |
| 9. | c | 10. | a | 11. | d | 12. | c |
| 13. | a | 14. | c | 15. | d | 16. | a |
| 17. | c | 18. | c | 19. | c | 20. | a |
| 21. | b | | | | | | |

# CHAPTER 15

# *Gametes, Fertilization, and Early Development*

## LEARNING OBJECTIVES

Be able to
1. Define the following terms:
    a. Gametogenesis
    b. Oogenesis
    c. Primordial germ cell
    d. Oocyte
    e. Lampbrush chromosome
    f. Animal pole
    g. Vegetal pole
2. Identify the role of each of the following in oogenesis:
    a. Vitellogenin
    b. Nurse cell
    c. Follicle cell
3. Describe the events of the
    a. Proliferation phase
    b. Growth phase
    c. Maturation phase
4. Define the following terms:
    a. Spermatogenesis
    b. Spermiogenesis
    c. Spermatid
5. Describe gametogenesis in a plant and differentiate between
    a. Sperm cell and pollen grain
    b. Embryo and endosperm
6. Describe the events of fertilization
    a. In an animal
    b. In a flowering plant
7. Define the following terms:
    a. Blastula
    b. Gastrula
    c. Differentiation
    d. Determination
    e. Morphogenesis
8. List four factors that influence differentiation.
9. Describe the role of each of the following in morphogenesis:
    a. Cell adhesion
    b. Cell migration
    c. Cell shape
    d. Morphogens
    e. Homeotic genes
10. Differentiate between programmed cell death and cell death associated with aging.
11. Briefly discuss the role of genetics and environmental factors in cell death.

## CHAPTER OVERVIEW

   I. GAMETOGENESIS, pp. 674-686
      1. Organisms produce haploid cells called gametes for sexual reproduction.
      2. Female gametes = eggs; male gametes = sperm.
      3. During gametogenesis, diploid primordial germ cells undergo meiosis and differentiation to become haploid egg and sperm cells.

# Chapter 15

**A. Oogenesis Is Delayed at Prophase I to Allow Time for Oocyte Growth, pp. 674-675**
1. In animals, egg formation is called oogenesis and takes place in the ovary.
2. During the proliferation phase, primordial germ cells (oogonia) divide by mitosis.
   a. In humans, proliferation is completed during embryonic development.
3. Now called oocytes, the cells go through meiosis.
   a. In nonmammalian oocytes, meiosis is delayed to allow an increase in cell size (growth phase).

**B. Lampbrush Chromosomes and Multiple Nucleoli Produce RNA That Is Stored in the Egg for Use after Fertilization, pp. 675-677**
1. During the growth phase, a high rate of transcription causes diplotene chromosomes to appear as lampbrushes (lampbrush chromosome).
2. The RNA produced is stored for RNP particles.
3. rRNA genes are amplified causing more nucleoli and ribosomes.

**C. Yolk Platelets Store Nutrients for Use by the Embryo after Fertilization, p. 677**
1. In birds, the liver produces vitellogenin for storage in the egg as yolk platelets.
2. Yolk platelets contain proteins, lipids, polysaccharides and hydrolytic enzymes. The enzymes are activated at fertilization to digest the yolk.
3. Follicle cells and oocytes are in close contact so nutrients can be transported from the follicle cells to the oocytes.
4. Mammals do not produce yolk platelets; nutrients are provided by follicle cells.
5. In invertebrates, nurse cells provide nutrients for oocytes.
6. In birds and amphibians, yolk proteins are synthesized on the rough ER of the oocyte.
7. The ER also produces pigment granules and cortical granules.

**D. Oocytes Develop into Asymmetric Cells That Acquire Various Kinds of Protective Coats, pp. 677-678**
1. Yolk platelets and nutrients are stored at the vegetal pole.
2. The nucleus, ribosomes, ER, mitochondria, and pigment granules are at the animal pole.
3. A polysaccharide coat develops on the oocytes. It is called the vitelline envelope (in nonmammalian oocytes) or zona pellucida (in mammalian oocytes).

|                | Produced by    | Notes                                   |
|----------------|----------------|-----------------------------------------|
| Primary coat   | Oocyte         |                                         |
| Secondary coat | Follicle cells |                                         |
| Tertiary coat  | Oviduct        | Egg white & external shell of chickens  |

**E. The Activation of MPF Triggers the Maturation Phase of Oogenesis, pp. 678-680**
1. Following the growth phase, the oocyte is arrested in prophase I until activated by a hormone to enter the maturation phase.
2. In amphibians: progesterone → $p39^{mos}$ → breakdown of cyclin inhibited → ↑[MPF] → metaphase I.
3. MPF causes the nuclear envelope to breakdown, spindle microtubules to assemble, and meiosis I.

## Chapter 15

  4. The first meiotic division is asymmetrical to produce a secondary oocyte and a polar body.
  5. The secondary oocyte goes through meiosis II to produce another polar body and an egg cell.
  6. In vertebrates, metaphase II doesn't occur until after fertilization.

**F. Spermatogenesis Begins with the Formation of Haploid Spermatids That Remain Connected to One Another by Cytoplasmic Bridges, pp. 680-681**
  1. Sperm cells are produced by spermatogenesis.
  2. In mammals, Sertoli cells provide protection and nutrition for developing sperm.
  3. Primordial germ cells (spermatogonia) divide by mitosis throughout the adult male's life.
  4. Some become primary spermatocytes and go through meiosis.
  5. Meiosis I produces secondary spermatocytes; meiosis II produces spermatids.
  6. The mitotic and meiotic spindles do not separate so spermatids remain connected.
  7. During spermiogenesis, spermatids develop into mature sperm.

**G. The Sperm Head Contains the Haploid Nucleus and an Acrosome, p. 681**
  1. During spermiogenesis
     a. Nonhistone proteins are removed.
     b. Normal histones are replaced.
     c. Chromatin is almost crystalline.
     d. DNA is metabolically inert.
  2. The nucleus localizes in the sperm head with the acrosome.
  3. The acrosome forms from the Golgi complex before the Golgi complex is expelled from the cell.
  4. The acrosome contains hydrolytic enzymes.

**H. The Sperm Tail Consists of a Flagellum and Associated Mitochondria, pp. 682-683**
  1. One centriole produces a flagellum.
  2. The axenome is 9 + 2 microtubules plus 9 accessory fibers.
  3. The mitochondria migrate to the middle piece that surrounds the base of the flagella.
  4. The remaining cytoplasm is discarded as the residual body.

**I. Gametogenesis in Flowering Plants Involves the Production of Eggs and Pollen Grains by the Ovary and Anther, pp. 683-686**
  1. Male gametogenesis occurs in the anthers.
     a. Diploid microsporocytes divide by meiosis to produce microspores.
     b. Microspores acquire a thick spore coat to form pollen grains.
     c. The haploid nucleus of a pollen grain divides mitotically to produce a tube cell and a generative cell.
     d. The generative cell divides to produce two sperm cells.
  2. Female gametogenesis occurs in the ovary.
     a. Diploid megasporocytes divide by meiosis to produce four megaspores (3 disintegrate).
     b. The megaspore divides mitotically to produce an embryo sac with eight haploid nuclei which mature to form 1 egg, 2 polar nuclei, 2 synergid cells, and 3 antipodal cells.

# Chapter 15

II. **FERTILIZATION, pp. 686-694**
   1. Sperm and egg fusion during fertilization which reestablishes a diploid cell.

   A. **Contact between Sperm and Egg Triggers the Acrosomal Reaction and Penetration of the Egg Coats, pp. 686-688**
      1. Sperm are propelled by flagella, cell crawling, or an aqueous environment.
      2. In sea urchins, the resact peptide of the jelly coat attracts sperm.
      3. After contact, $Ca^{2+}$ channels in the sperm open which activates an $Na^+$-$H^+$ antiport; the internal pH rises causing the acrosomal reaction.
      4. In the acrosomal reaction, the acrosome fuses with the sperm plasma membrane and expels its hydrolytic enzymes to digest the outer coating of the egg.
      5. Polymerization of actin molecules forms the acrosomal process and bindin binds receptors on the vitelline envelope.

   B. **Following the Fusion of Sperm and Egg, a Transient Depolarization of the Plasma Membrane Prevents Polyspermy, p. 688**

   C. **An Increased $Ca^{2+}$ Concentration in the Egg Cytosol Triggers the Cortical Reaction, Leading to a Permanent Block to Polyspermy, pp. 688-691**
      1. During depolarization, the cortical reaction occurs.
      2. Cortical granules fuse with the plasma membrane and discharge their contents between the membrane and the vitelline envelope.
      3. Enzymes digest sperm receptors and change the vitelline envelope into a rigid fertilization membrane.
      4. The cortical reaction is triggered by $Ca^{2+}$. $Ca^{2+}$ is released from the ER mediated by the second messenger $IP_3$.

   D. **Movements of the Sperm Nucleus and the Egg Cytoplasm Occur Shortly after Fertilization, p. 691**
      1. When the egg cytoplasm makes contact with the sperm, the fertilization cone draws the sperm nucleus into the egg.
      2. A new membrane forms around the sperm nucleus to create the male pronucleus.
      3. Microtubules move the male pronucleus toward the female pronucleus; the pronuclei fuse.
      4. The sperm centrioles will produce the centrosomes that will form the mitotic spindle.
      5. The outer cytoplasm of the egg rotates about 30°C relative to the inner cytoplasm.

   E. **An Increase in the pH of the Fertilized Egg Activates Protein Synthesis and DNA Replication, pp. 691-692**
      1. mRNA present in the unfertilized egg is activated and translated following fertilization.
      2. The intracellular pH of an unfertilized egg is 6.8 which keeps the cell inactive.
      3. When the pH is raised, protein synthesis and DNA replication begin.
      4. After fertilization, the $Na^+$-$H^+$ antiport pumps $H^+$ out of the cell.

# Chapter 15

    **F. Two Fertilization Events Occur in Flowering Plants, Producing a Diploid Zygote and Triploid Endosperm, pp. 692-694**
1. When pollen makes contact with proteins on the stigma of the appropriate species, the pollen tube grows.
2. In the embryo sac, the pollen tube penetrates one of the synergids and releases the two sperm cells.
3. One sperm cell fuses with the egg to form a diploid zygote; the other fuses with the two polar nuclei to produce a triploid nucleus.
4. The triploid nucleus will form the (nutritive) endosperm.
5. The zygote and endosperm divide mitotically.
6. Growth is arrested to form the seed consisting of the embryo surrounded by the endosperm.

## III. FORMATION OF THE EMBRYO, pp. 694-711

    **A. Cleavage Converts the Fertilized Egg into a Multicellular Blastula, pp. 695-696**
1. Animal eggs can be fertilized in a variety of stages depending on the species.
2. Amphibian eggs are arrested in metaphase II by $p39^{mos}$, fertilization causes an increase in $Ca^{2+}$.
3. $Ca^{2+}$-dependent protease (calpain) degrades $p39^{mos}$.
4. Following meiosis, rapid division without growth occurs to produce a hollow ball of cells called a blastula.

    **B. Gastrulation Converts a Blastula into a Three-Layered Gastrula Composed of Ectoderm, Mesoderm, and Endoderm, p. 696**
1. An opening forms in the blastula called the blastopore.
2. Three cell layers form and the embryo is called a gastrula:
   a. Ectoderm—develops into epithelia covering external surfaces and neural tissue.
   b. Endoderm—develops into epithelia covering internal tissues.
   c. Mesoderm—develops into blood, bone, and connective tissue.
3. Embryonic cells differentiate into specialized cells.
4. Early in cleavage, cells are capable of becoming many different types of cells; when cells become committed it is called determination.

    **C. Cytoplasmic Segregation of Morphogenic Determinants Can Influence How Cells Differentiate, pp. 696-700**
1. Morphogenic determinants are asymmetrically distributed in eggs; they influence cell differentiation. For example,
   a. In snails, the polar lobe influences differentiation.
   b. In amphibians, the gray crescent influences cell differentiation.
2. The kind of cytoplasm (e.g., animal pole or vegetal pole) can influence differentiation.
3. Cytoplasmic granules called polar granules in *Drosophila* correlate with the formation of germ cells.
4. Cytoplasmic granules in *Caenorhabditis elegans* correlate with the formation of sperm and egg cells.

## Chapter 15

D. **Inductive Interactions between Neighboring Cells Can Also Influence How Cells Differentiate, pp. 700-701**
   1. Induction occurs when early gastrula cells destined to become nervous tissue transplanted to another embryo in a region destined to become epidermis develop into epidermis.
   2. Primary embryonic induction occurs when an entire new embryo is produced by transplanted cells.
   3. Secondary induction occurs when one group of cells causes a neighboring group of cells to differentiate into a particular structure.

E. **Embryonic Inductions are Mediated by Cell-Cell Contact, Cell-Matrix Contact, and Diffusion of Signaling Molecules, pp. 701-702**
   1. Induction of glandular tissues requires cell-cell contact between inducing and responding cells.
   2. Induction of the cornea requires contact between the inducing matrix and the responding cells.
   3. Induction of neural tissue requires nonspecific signaling molecules.

F. **Cell Differentiation Is Accomplished through Changes in Gene Expression, pp. 702-703**
   1. Nuclei from differentiated cells that are transplanted into enucleated eggs will cause the egg to develop.
   2. These nuclei are totipotent, that is, possess all the genes required for the development of a new organism.
   3. Differentiated plant cells are totipotent. Single cells from carrot roots will grow into complete plants in culture media.
   4. Protoplasts will grow into a clump of undifferentiated cells (a callus) which eventually develops shoots and roots.

G. **Gene Expression in Differentiated Cells Is Controlled by a Variety of Molecular Mechanisms, pp. 703-704**
   1. Housekeeping genes are expressed in all the cells of an organism.
   2. Tissue-specific genes which are expressed in certain cell types are controlled by:
      a. Gene-specific transcription factors
      b. Different splicing of a particular RNA will produce different mRNAs

H. **Changes in Gene Expression Can Determine the Developmental Fate of Differentiating Cells, p. 704**
   1. Some gene expression is the result of differentiation. E.g., globin is produced after differentiation into red blood cells.
   2. Some gene expression causes differentiation. E.g.,
      a. The *lin-12* gene in *Caenorhabditis*.
      b. The *Notch* gene in *Drosophila*.
      c. The *MyoD* gene in vertebrate muscle cells.

I. **Cell-Cell Recognition and Adhesion Play a Central Role in Morphogenesis, pp. 705-706**
   1. The physical arranging of differentiating cells into a multicellular pattern is called morphogenesis.
   2. In a mixture of cell types, like cells aggregate internal to the next cell type in the list: epidermis, cartilage, pigmented epithelium, myocardium, nervous tissue, liver.

3. Adhesion molecules such as N-CAM are responsible for morphogenesis.
4. Contact between adjacent cells might alter membrane receptors to influence cells.
5. Contact between cells may form a gap junction to permit electrical/chemical signals.

J. **Cell Migration and Changes in Cell Shape Also Contribute to Morphogenesis, pp. 706-707**
1. Interactions between integrins and extracellular matrix components are responsible for cell migration.
2. Tissue invagination and evagination may be influenced by cell shape. Cell shape is controlled by actin filaments.

K. **Morphogen Gradients Allow Cells To Locate Their Positions within the Developing Embryo, p. 707**
1. The concentration of diffusible molecules called morphogens allows cells to sense their location.

L. **Homeotic Genes Control Development in Different Regions of the Embryo, pp. 708-709**

M. **Programmed Cell Death Is a Normal Part of Morphogenesis, p. 709**
1. Programmed cell death leads to the destruction of specifically targeted cells such as the tissue between developing fingers and toes.
2. During apoptosis cells rapidly die and shed membrane vesicles which are ingested and destroyed by neighboring cells.

N. **Genetic Factors and Oxidative Damage Contribute to Cell Death Associated with Aging, pp. 709-711**
1. The number of cells divisions possible is genetically determined.
2. Cells can be damaged and their lifespan shortened by free radicals.
3. Vitamin E blocks the action of free radicals.
4. Superoxide dismutase destroys superoxide free radical.
5. Aging and cell death could be related to the shortening of the telomere during DNA replication.

# KEY TERMS

| | | |
|---|---|---|
| acrosomal process | acrosomal reaction | acrosome |
| activin | aequorin | animal pole |
| anther | antipodal | apoptosis |
| bindin | blastopore | blastula |
| calpain | cleavage | cortical granule |
| cortical reaction | determination | differentiation |
| ectoderm | egg cell | embryo sac |
| embryo sac | endoderm | endosperm |
| epithelia | female pronucleus | fertilization |
| fertilization membrane | follicle cell | free radicals |
| gamete | gametogenesis | gastrula |
| generative cell | gray crescent | head |
| homeotic genes | housekeeping genes | induction |
| lampbrush chromosome | male pronucleus | megaspore |
| megasporocyte | mesoderm | microspore |

# Chapter 15

- microsporocyte
- morphogens
- ovary
- polar body
- pollen tube
- primordial germ cell
- secondary induction
- sperm cell
- spermatogenesis
- synergid
- tube cell
- vitellogenin
- zone of polarizing activity
- morphogenesis
- nurse cell
- p39$^{mos}$
- polar lobe
- polyspermy
- resact
- seed
- sperm tail
- spermiogenesis
- tissue-specific genes
- vegetal pole
- yolk platelets
- zygote
- morphogenic determinants
- oocyte
- pigment granule
- pollen
- primary embryonic induction
- retinoic acid
- Sertoli cell
- spermatid
- stigma
- transcription unit
- vitelline envelope
- zona pellucida

## KEY FIGURE

Label the sperm nucleus, acrosome, egg plasma membrane, cortical granule, vitelline envelope, actin, acrosomal process, and fertilization membrane.

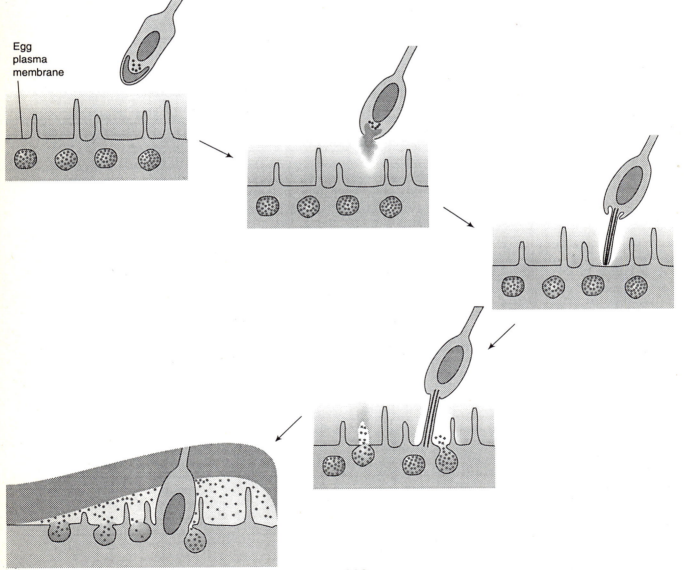

# Chapter 15

## 🖝 STUDY QUESTIONS

1. During the proliferation phase, oogonia
    a. Become haploid cells.
    b. Increase in number.
    c. Divide by meiosis.
    d. Differentiate into eggs and sperm.
    e. None of the above.

2. Most nutrients for human oocytes are provided by
    a. The liver.
    b. Follicle cells.
    c. Nurse cells.
    d. The bloodstream.
    e. None of the above.

3. The following events are required to trigger metaphase I in an oocyte. Which is the fourth step?
    a. Breakdown of cyclin inhibited
    b. Metaphase I
    c. $p39^{mos}$ produced
    d. Progesterone is produced
    e. ↑[MPF]

Use the following choices to answer questions 4-5.
    a. Growth
    b. Maturation
    c. Proliferation
    d. All of the above
    e. None of the above

4. During which phase is the first meiotic division completed?

5. During which phase are yolk platelets acquired?

For questions 6-7: Compare the paired choices for each question and answer
    a  if A is greater than B
    b  if B is greater than A
    c  if the two are nearly equal

> For example:  The ease of seeing bacteria
>     A.   With a microscope
>     B.   Without a microscope          Answer:   a

6. Mitochondria in
    A. An egg
    B. A sperm

7. Transcription in
    A. An egg
    B. A sperm

# Chapter 15

8. The following events occur in plant male gametogenesis. What is the third step?
   a. A tube cell and generative cell forms.
   b. Microspores develop into pollen grains.
   c. Microsporocytes divide by meiosis.
   d. Sperm cells develop.

9. In humans, fertilization could be inhibited by blocking all of the following **except**
   a. Bindin receptors.
   b. Actin polymerization.
   c. $Ca^{2+}$ channels.
   d. Acrosomal enzyme.
   e. Resact.

Use the following figure to answer questions 10-11. Diploid nuclei are shaded circles; haploid nuclei are white circles.

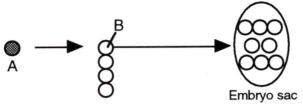

10. Cell A is a(n)
    a. Egg cell.
    b. Megasporocyte.
    c. Microcyte.
    d. Antipodal cell.
    e. None of the above.

11. To produce the embryo sac, cell B undergoes
    a. Mitosis.
    b. Meiosis.
    c. Differentiation.
    d. Induction.
    e. Cleavage.

12. Protein synthesis in a zygote is not blocked by inhibitors of RNA synthesis indicating that
    a. The mRNA is already present in the unfertilized egg.
    b. Eggs make a different type of mRNA.
    c. Protein synthesis is activated by a pH increase.
    d. DNA is being replicated.
    e. None of the above.

13. Which one of the following is **not** correctly matched?
    a. Ectoderm—develops into epithelia covering external surfaces
    b. Ectoderm—develops into neural tissue
    c. Endoderm—develops into the digestive system
    d. Mesoderm—develops into blood, bone, and connective tissue
    e. None of the above

## Chapter 15

14. A cell from the root of a plant can grow into a complete plant. This indicates that in differentiated cells
    a. Unneeded DNA is destroyed.
    b. DNA is destroyed during cell differentiation.
    c. Each type of cell in an organism has only the information it needs for particular functions.
    d. Each type of cell in an organism has all the organism's DNA.
    e. None of the above.

15. Place the following events of fertilization in order:
    1-Bindin penetrate the egg's glycocalyx.
    2-The egg's plasma membrane undergoes a change in membrane potential.
    3-The sperm and egg nuclei fuse.
    4-DNA replication begins.
    5-The acrosomal reaction releases hydrolytic enzymes from the acrosomal vesicle.
    6-Calcium is released from egg cell storage sites.
    7-The first cell division occurs.

    a. 1,2,3,4,5,6,7
    b. 1,5,2,6,3,4,7
    c. 1,3,2,5,4,6,7
    d. 1,5,6,4,3,2,7

Use the following choices to answer questions 16-18. Choices may be used once, more than once, or not at all.
    a. Cytoplasmic determinants
    b. Morphogenesis
    c. Determination
    d. Induction
    e. None of the above

16. Early gastrula cells destined to become nervous tissue transplanted to another embryo in a region destined to become epidermis will develop into epidermis. This is an example of _____.

17. In a developing frog embryo, one of the capsules surrounding the developing eye is removed and the overlaying epidermis does not produce a cornea. This illustrates _____.

18. If the nucleus from an adult frog cell is transplanted into an enucleated frog egg, full development of the egg into a normal adult occurs. This illustrates _____.

Use this information to answer questions 19-20. Hepatocytes normally produce albumin. Albumin production is inhibited when hepatocytes are fused with intact fibroblasts or with enucleated fibroblasts.

19. Which one of the following is **true?**
    a. Albumin is tissue-specific.
    b. Albumin is produced by housekeeping genes.
    c. Fibroblast nuclei are totipotent.
    d. Hepatocytes are totipotent.
    e. None of the above.

20. Mesoderm cells develop into muscle cells when the *MyoD* gene is activated. If *MyoD* is inserted into liver cells, you would expect
    a. The gene will be destroyed.
    b. Mesoderm will develop.
    c. Liver cells will produce muscle-specific proteins.
    d. Muscle cells will start producing liver-specific proteins.
    e. None of the above.

21. In humans, B cells produce large quantities of the proteins called antibodies. These same cells do not make keratin or globulin. One would expect that B cells have
    a. Only genes for antibodies
    b. The genes for globulin, keratin, and antibodies
    c. Antibody genes and a few other genes
    d. Only tissue-specific genes
    e. Fewer genes than other cells

## PROBLEMS

1. In flowering plants a mature embryo sac with its surrounding tissue provides an ideal source of material for the study of the relationship between genetic expression and ploidy. Considering an embryo sac just after double fertilization, explain how such a study might be conducted and outline the variety of genetic conditions that can be studied.

2. Germination takes place after an appropriate period of dormancy in a typical dicot seed. What must happen to the seed before germination takes place? What is the fate of the endosperm? As cell division occurs, the top cells secrete auxin that diffuses down. Describe the resulting growth pattern.

3. Sexual dimorphism, that is, differences between males and females of a species, is often created by cell death. The songs of male and female zebra finches are different. The song is controlled by a site in the brain called the RA nucleus. Estrogen treatment during development causes a male song in female birds. Explain the role of cell death in song production.

4. Homeotic genes have been extensively studied in *Drosophila embryos*. The embryo is 14 parasegments long, with parasegment 1 in the head region. The products of *Antp* are normally present in parasegments 4 and 5 and the products of *Ubx* are found in parasegments 5-6. If *Ubx* is deleted, products of *Antp* are found in parasegments 5-6. If gene *abd* is deleted, *Antp* and *Ubx* products are found through parasegment 12. Since these genes are present in every cell in the fly's body, propose a mechanism for controlling their expression.

5. In the embryo shown below, the cells marked A will produce neural tissue. The cells marked B will produce intestine. If the A pair is isolated and cultured, only ectoderm results. What do you expect if pair A and pair B are cultured together? Explain the results.

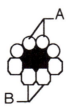

# Chapter 15

6. If an egg from a 32-year-old woman is fertilized, is the zygote 32 years old? Discuss your answer.

7. The site of sperm penetration of the egg will become the ventral surface of the organism. Describe the events that make this happen.

## ☞ ANSWERS TO STUDY QUESTIONS

| | | | | | | | |
|---|---|---|---|---|---|---|---|
| 1. | b | 2. | b | 3. | e | 4. | b |
| 5. | a | 6. | a | 7. | a | 8. | a |
| 9. | e | 10. | b | 11. | a | 12. | a |
| 13. | c | 14. | d | 15. | b | 16. | d |
| 17. | d | 18. | a | 19. | a | 20. | c |
| 21. | b | | | | | | |

# CHAPTER 16

## Lymphocytes and the Immune Response

### ⇧ LEARNING OBJECTIVES

Be able to
1. Differentiate between the following terms:
   a. Antigen and antibody
   b. Antigen and epitope
   c. B cell and T cell
   d. T cell and natural killer cell
   e. MHC I and MHC II
   f. Plasma cell and memory cell
2. Describe how antigens are processed by antigen-presenting cells.
3. Describe the structure of an antibody.
4. Explain the clonal selection theory.
5. Identify the contributions made by Landsteiner, Jerne and Burnet, and Tonegawa.
6. Explain how V, C, D, and J gene segments result in antibody diversity.
7. Describe how antigen-presenting cells, T cells, and B cells interact to elicit antibody formation.
8. List the effects of complement activation.
9. Differentiate between the following terms:
   a. Lymphocyte and T cell
   b. Helper T cell and killer T cell
   c. Killer T cell and NK cell
10. Define lymphokine and identify the functions of the following:
    a. Interleukin-1
    b. Interleukin-2
11. Define interferon and explain its method of action.
12. Describe how immunological tolerance develops according to the deletion theory and anergy.
13. Describe how HIV causes an immune deficiency.

### ⇧ CHAPTER OVERVIEW

   I. **INTRODUCTION, p. 715**
      1. Specialized white blood cells called lymphocytes secrete antibodies that bind to specific molecules and kill cells directly.

  II. **OVERVIEW OF THE IMMUNE SYSTEM, pp. 715-718**

      A. **Antigens Are Substances That Elicit an Immune Response, p. 715**
         1. An antigen is a molecule capable of eliciting an immune response.
         2. The more foreign an antigen is, the greater will be the immune response.
         3. Epitopes are the molecular sites on an antigen that evoke the immune response.

      B. **Antigens Are Processed by Antigen-Presenting Cells, pp. 715-717**
         1. Macrophages phagocytize foreign cells and particles.
         2. Macrophages that carry MHC class II molecules can be antigen-presenting cells.
         3. To process exogenous antigens:
            a. Antigens are brought into the cell.

b. Antigens are digested by lysosomal enzymes.
c. Antigen fragments bind to MHC molecules.
d. The antigen-MHC complex is transported to the cell surface.
4. To process endogenous antigens:
a. Antigens (e.g., viral proteins) originating inside a cell bind to MHC class I molecules in the ER lumen of any nucleated cell.
b. The antigen-MHC complex is transported to the cell surface.

**C. B Cells Produce Antibodies and T Cells Are Responsible for Cell-Mediated Immunity, pp. 717-718**
1. B cells mature in the bone marrow and migrate to secondary lymphoid organs, e.g., lymph nodes and spleen.
2. T cells mature in the thymus gland and migrate to secondary lymphoid organs.
3. B cells produce antibodies.
4. T cells kill cells in cell-mediated immunity.
5. Natural killer (NK) cells kill tumor cells and virus-infected cells.

## III. ANTIBODY STRUCTURE AND DIVERSITY, pp. 718-727

**A. Antibodies Are Constructed from Light and Heavy Chains, pp. 718-720**
1. Antibodies or immunoglobulins are proteins synthesized by B cells in response to an antigen then secreted into body fluids.

| Antibody | Heavy chains | Notes |
|---|---|---|
| IgG | $\gamma$ | Blood |
| IgA | $\alpha$ | Secretions |
| IgM | $\mu$ | Blood |
| IgD | $\delta$ | Surfaces of B cells |
| IgE | $\epsilon$ | Parasitic infections and allergic reactions |

2. Antibodies can be digested with papain to yield three fragments.
a. The two Fab fragments can combine with antigen.
b. The Fc region does not react with antigens.
3. Breaking the disulfide bonds in antibodies yields two heavy chains and two light chains.
4. Each of the classes of antibodies has a different type of heavy chain.
5. Light chains will be either $\kappa$ or $\lambda$.
6. In IgA and IgM antibodies, monomers are held together by a polypeptide called the J chain.

**B. Light Chains and Heavy Chains Are Composed of Constant and Variable Regions, p. 720**
1. Antibodies were first studied in multiple myeloma patients because they produce IgG antibodies in excess.
2. All light chains of the same type ($\kappa$ or $\lambda$) have the same constant region and all heavy chains of the same type ($\alpha, \gamma, \mu, \delta,$ or $\epsilon$) have the same constant region.
3. The constant region of the IgG heavy chain is comprised of three domains of 100 amino acids each.

**C. Antigen-binding Sites Are Formed by the Hypervariable Regions of Heavy and Light Chains, pp. 720-722**
1. Differences in antibodies are due to differences in three hypervariable regions of 5-15 amino acids.

# Chapter 16

    2. Folding of the heavy and light chains forms the antigen-binding site from the hypervariable regions.

**D. Clonal Selection Allows Specific Antibodies To Be Formed in Response to Each Antigen, p. 722**
1. A hapten is a molecule that will elicit an antibody response when joined to a larger carrier molecule. The hapten alone can combine with the antibodies.
2. Landsteiner discovered that antibodies are specific for an antigen.
3. Pauling suggested the instructive theory of antibody formation which states that antigens interact with antibodies as they are being synthesized.
4. Jerne and Burnet proposed the now accepted clonal selection theory which states that a human possesses millions of different lymphocytes, each capable of producing a different antibody.
5. Each antibody-producing cell makes a single type of antibody and a particular cell is selected by the presence of an antigen.

**E. DNA Segments Coding for the Variable and Constant Regions of Antibody Molecules Are Rearranged in Antibody-Producing Cells, pp. 722-723**
1. If each antibody were encoded by a different gene, all of an organism's DNA would be taken up to meet this requirement.
2. The germ line theory states that ~2,000 genes for light and heavy chains are inherited and different antibodies are made by varying combinations of heavy and light chains.
3. Alternatively, an organism could have a few antibody genes that recombine or mutate during lymphocyte development.
4. Tonegawa discovered that C gene segments code for the constant regions of antibodies and V gene segments code for the variable regions of antibodies.
5. C and V genes are rearranged during embryonic development.

**F. Recombination Involving Multiple V, D, and J Segments Is a Major Contributor to Antibody Diversity, pp. 723-724**
1. Human DNA contains ~80 different V segments for the light chain and 100-200 V segments for heavy chains.
2. 10-15 amino acids at the end of the variable region are coded for by J gene segments.
3. Part of the variable region of a heavy chain is encoded by D gene segments

| Antibody | C | Number of gene segments V | J | D |
|---|---|---|---|---|
| Heavy chain | $\alpha, \delta, \gamma, \mu,$ and $\varepsilon$ | 100-200 | 6 | 20 |
| Light chain | $\kappa$ or $\lambda$ | 80 | | |
| $\kappa$ | | | 5 | |
| $\lambda$ | | | 4 | |

4. During B cell development, any of the V genes can be spliced to any J and any D gene. This would give $200 \times 6 \times 20 = 24,000$ different heavy chains. $80 \times 5 = 400$ different $\kappa$ chains; $80 \times 4 = 320$ $\lambda$ chains.
5. Heavy and light chains combine randomly to give $24,000 \times 400 = 9.6 \times 10^6$ different antibodies with $\kappa$ chains; or $24,000 \times 320 = 7.68 \times 10^6$ with $\lambda$ chains.

**G. Antibody Diversity Is Further Increased by Inexact Joining of Gene Segments and Somatic Mutations, pp. 724-726**
1. R sequences occur at the 3′ end of each V segment, on both sides of each D segment, and at the 5′ end of each J segment.

# Chapter 16

    2. The complementary R sequences form DNA loops that bring DNA sequences together.
    3. The position of a splice may vary by several nucleotides.
    4. The D and J segments code for hypervariable region CDR3.
    5. The rate of mutations in hypervariable regions CDR1 and CDR2 is 1 million times higher than other DNA sites.

**H. All Antibodies Produced by a Given B Cell Have the Same Antigen-Binding Site, p. 726**
1. A B cell produces one V-J-D segment coding for a heavy chain and one V-J segment coding for a light chain.
2. Transcription produces an RNA transcript containing extra bases between the variable and constant regions as well as introns.
3. The extra bases and introns are removed during RNA processing.

**I. Secreted and Membrane-Bound Forms of the Same Antibody Are Made by Altering the RNA Transcribed from the Heavy Chain Gene, pp. 726-727**
1. Antibodies bound to the plasma membrane function as antigen receptors.
2. After a B cell is selected, the secreted form of the antibody is produced.

## IV. CELL INTERACTIONS DURING ANTIBODY FORMATION, pp. 727-735

**A. Antibodies Are Made by B Cells with Assistance from Helper T Cells, pp. 727-728**
1. Removal of the thymus in newborn mice decreases their ability to produce antibodies.
2. Newborn chicks whose bursa of Fabricius has been removed cannot make antibodies. (In chickens, B cells mature in the bursa of Fabricius.)

**B. Helper T Cells Bind to Antigen-MHC II Complexes on the Surface of Antigen-Presenting Cells, pp. 728-729**
1. Helper T cells have T-cell antigen receptors (TCR) which bind to antigen on an antigen-presenting cell.
    a. Alpha-beta TCRs recognize foreign peptides; gamma-delta TCRs recognize damaged host tissues.
    b. TCRs are produced from V, J, and D gene segments that randomly recombine during T-cell development.
    c. Each helper T cell makes a single type of TCR.
2. Helper T cells have CD4, a surface glycoprotein that binds to MHC II on an antigen-presenting cell.

**C. Interleukins Stimulate the Proliferation of Helper T Cells, pp. 729-730**
1. Lymphokines or cytokines are proteins secreted by cells that influence growth and differentiation in cells.
2. The antigen-presenting cell secretes interleukin-1 when it binds to a helper T cell. The helper T cell secretes IL-2.
3. IL-1 causes bound T cells to release IL-2; IL-2 causes proliferation of T cells.

## Chapter 16

D. **B Cells Are Activated by Helper T Cells through a Mechanism that Also Involves Interleukins, pp. 730-732**
1. Each B cell has 20,000 to 200,000 surface antibody molecules which bind to antigens.
2. A bound antigen is taken in by the B cell, digested, and returned to the cell surface bound to MHC II.
3. A helper T cell binds the antigen-MHC II complex and secretes IL-2, IL-4, IL-5, and IL-6,
4. The B cell divides.

E. **Proliferating B Cells Develop into Plasma Cells and Memory Cells, p. 732**
1. Antibody secreting cells are called plasma cells.
2. IgM antibodies are made first, then IgG, IgA, or IgE. This is called class switching.
3. Some B cells become memory B cells; they do not secrete antibodies.
4. Memory B cells divide and become plasma cells on a second exposure to the same antigen (secondary immune response).
5. The secondary immune response is more rapid and produces more antibodies than the primary immune response.

F. **The Complement System Assists Antibodies in Defending Organisms against Bacterial Infections, pp. 732-735**
1. Antibodies can bind to a soluble antigen to block its action.
2. The complement system is a group of ~20 circulating proteins.
3. In the classical pathway, complement is activated by an antigen-antibody reaction. In this example, assume the antigen is on a bacterial cell. Although complement proteins are numbered, they don't react exactly in numerical order!
    a. **C1** binds to the heavy chain of the constant region.
    b. **C4, C2,** and **C3** are cleaved.
    c. C4b, C2b, and C3b combine to activate **C5 through C9.**
    d. C9 forms the membrane-attack complex which forms pores in the cell membrane. Small molecules leak out of the cell, leading to osmotic lysis of the cell.
4. In the alternative pathway, **C3b** and factor B bind to gram-negative bacterial cell walls; **C5 through C9** are activated to form the membrane-attack complex.
5. C3b binds to receptors on macrophages and neutrophils to promote phagocytosis.
6. C3a stimulates a local inflammatory response by binding to mast cells which release histamine and phagocyte chemotactic factor.

V. **CELL-MEDIATED IMMUNITY, pp. 735-739**
1. Some parasites can evade the immune system by living inside host cells.
2. In cell-mediated immunity, lymphocytes attack infected cells.

A. **Killer T Cells Are Responsible for Cell-Mediated Immunity, pp. 735-736**
1. Cytotoxic T cells or killer T cells possess CD8 on their surfaces.
2. CD8 is the TCR.

## Chapter 16

- B. **Killer T Cells Bind to Antigen-MHC I Complexes on Target Cells, pp. 736-737**
    1. Virus-infected cells often produce viral proteins associated with MHC I on their surface.
    2. TCRs on killer T cells recognize the antigen-MHC I complex.
    3. A CD8 cell bound to a target cell is called a pre-killer cell.
    4. Helper T cells secrete interleukins that cause the pre-killer cells to divide and form a clone of killer cells and memory T cells.

- C. **Killer T Cells Destroy Their Target Cells by Secreting the Pore-Forming Protein, Perforin, p. 737**
    1. Perforin stored in vesicles is secreted from killer T cells.
    2. Perforin inserts into the plasma membrane of target cells and polymerizes into a doughnut shape to form a pore.

- D. **Grafted Tissues Are Rejected by Cell-Mediated Immune Responses Directed against Foreign MHC Molecules, pp. 737-738**

- E. **NK Cells Destroy Cancer Cells by a Mechanism That Does Not Involve the Recognition of Antigen-MHC Complexes, pp. 738-739**

- F. **Interferons Provide an Early Nonspecific Immunity to Viral Infections, p. 739**
    1. Interferons secreted by virus-infected cells enter neighboring cells to cause the production of antiviral enzymes.

VI. **TOLERANCE, AUTOIMMUNITY, AND AIDS, pp. 739-745**

- A. **Tolerance Helps the Immune System Distinguish between Self and Nonself, pp. 739-741**
    1. Antigens encountered during embryonic development are recognized as self and induce immunological tolerance.
    2. According to the deletion theory, T cells directed against self are destroyed during embryonic development.
    3. The deletion theory was tested using transgenic SCID mice that produce only one type of TCR.
    4. In the thymus, immature T cells carrying a TCR directed against self were destroyed.
    5. Immature T cells that are not destroyed and that bind to MHC I lose their CD4 and develop into CD8 cells; immature T cells that bind to MHC II lose their CD8 and develop into CD4 cells.
    6. Tolerance induced by low doses of antigen is caused by suppressor T cells.
    7. When cells bind to antigen and become inactivated, anergy results. Anergy is prevented by mixing antigen with an adjuvant (additive that stimulates the immune response).

- B. **The Failure of Immunological Tolerance Can Trigger Autoimmune Diseases, pp. 741-742**
    1. In autoimmune diseases, the immune response is directed against self antigens.
    2. Mice injected with inactivated T cells against self produce antibodies against the TCRs on the T cells. (Antibodies, e.g., TCRs, acting as antigens are called idiotypes. The mouse made anti-idiotype T cells in response to the injected T cells.)

## Chapter 16

C. **Acquired Immunodeficiency Syndrome (AIDS) Is Caused by a Depletion of Helper T Cells Associated with HIV Infection, pp. 742-745**
1. AIDS is caused by Human Immunodeficiency Virus (HIV).
2. HIV infects helper T cells and the helper T cells die.
3. HIV is an RNA virus with reverse transcriptase.
4. To reproduce:
   a. HIV gp120 binds to CD4.
   b. The viral envelope fuses with the plasma membrane.
   c. The virus is uncoated and viral DNA is produced.
   d. Viral DNA integrates in the host chromosome and is transcribed to produce viral proteins.
   e. New viruses bud from the cell.
5. The host produces antibodies and cell-mediated immunity against HIV, however, the virus mutates rapidly.
6. AZT is incorporated into viral DNA in place of deoxythymidine; AZT prevents elongation of the DNA.
7. Mutations in reverse transcriptase have led to resistance to AZT.

## KEY TERMS

| | | |
|---|---|---|
| AIDS | allograft | alternative pathway |
| anergy | antibody | antigen |
| antigen-presenting cell | autoimmune disease | B cell |
| C gene segment | CD4 | CD8 |
| cell-mediated immunity | class switching | classical pathway |
| clonal selection | complement system | constant region |
| D gene segment | deletion theory | epitope |
| helper T cell | HIV | hybridoma |
| hypervariable region | idiotype | immunoglobulin |
| interferon | interleukins | J gene segment |
| lymphocyte | lymphokines | macrophage |
| major histocompatibility complex | membrane-attack complex | memory B cell |
| memory T cell | MHC class I | MHC class II |
| natural killer cell | nude mice | perforin |
| plasma cell | R sequences | suppressor T cell |
| T cell | T-cell antigen receptor | tolerance |
| V gene segment | variable region | xenograft |

Chapter 16

## 👉 KEY FIGURE

Identify the T cell, B cell, T-cell clones. and B-cell clones. Label the TCR, MHC II, CD4, and antibody. Show the paths of IL-1 and IL-2. What is in the phagosome? _____

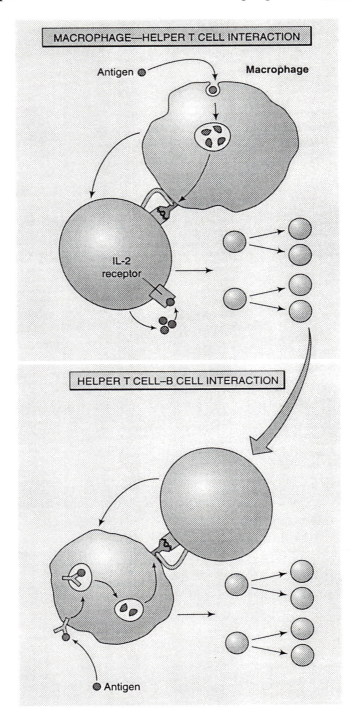

222

# Chapter 16

## ☞ STUDY QUESTIONS

Use the following choices to answer questions 1-4.
    a. APC
    b. Plasma cells
    c. Helper T cells
    d. Killer T cells
    e. Natural killer cells

1. Cells that produce antibodies.

2. Cells that kill virus-infected cells without antibodies.

3. Cells that phagocytize antigens.

4. Cells that bind to antigen-MHC II complex.

Use this figure to answer questions 5-7.

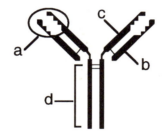

5. The Fc region.

6. In IgG, c is
    a. An α chain.
    b. A δ chain.
    c. A γ chain.
    d. A λ chain.
    e. A μ chain.

7. How many molecules would result if the molecule shown is treated with mercaptoethanol?
    a. 1
    b. 2
    c. 4
    d. 8
    e. None of the above

Use the figure below to answer questions 8-9.

| $V_1$ | $V_2$ | $V_3$ | $V_4$ | $V_5$ | $D_1$ | $D_2$ | $D_3$ | $D_4$ | $D_5$ | $D_6$ | $J_1$ | $J_2$ | $J_3$ | $J_4$ | $C_\mu$ | $C_\delta$ | $C_\gamma$ | $C_\varepsilon$ | $C_\alpha$ |
|---|---|---|---|---|---|---|---|---|---|---|---|---|---|---|---|---|---|---|---|

8. How many different types of IgG molecules can be made from these gene segments?
    a. 0
    b. 20
    c. 120
    d. 600
    e. Too many to count

9. How many different types of light chains can be made from the gene segments shown?
   a. 0
   b. 20
   c. 120
   d. 600
   e. Too many to count

10. The difference between antigen receptors on a B cell and antibodies secreted by the same B cell is
    a. The antigen-binding sites.
    b. The C-terminal region.
    c. The light chains.
    d. The heavy chains.
    e. The D segment.

11. Which one of the following best describes the immune response?
    a. An individual acquires all his antibodies from the mother's circulation.
    b. Genes for antibodies are inherited from both parents.
    c. Antibody genes are produced during embryonic development.
    d. Antigens interact with antibodies during their formation.
    e. Antibody genes form after exposure to an antigen.

12. CD4 differs from CD8 in that CD4
    a. Is associated with T cells.
    b. Is a TCR.
    c. Binds to antigen in solution.
    d. Binds to antigen-MHC II complexes.
    e. Is associated with killer T cells.

13. Killer T cells are responsible for destroying all of the following except
    a. Bacteria.
    b. Allografts.
    c. Xenografts.
    d. Viral-infected cells.
    e. None of the above.

14. Interferons are produced by ___1___ to ___2___.
    a. 1-Healthy cells; 2-protect them
    b. 1-Infected cells; 2-protect neighboring cells
    c. 1-Infected cells; 2-cure themselves
    d. 1-Healthy cells; 2-kill viruses
    e. None of the above

15. If you examined the blood of a patient with AIDS, which one of the following would you find?
    a. Antibodies
    b. No antibodies
    c. Too many suppressor T cells
    d. No killer T cells
    e. None of the above

## Chapter 16

16. In a patient with a gram-negative bacterial infection
    a. C3B activates C5 through C9.
    b. C1 activates C4, C2, C3, and C5 through C9.
    c. C4, C2, and C3 are cleaved.
    d. C3 is cleaved.
    e. None of the above.

17. A person does not make antibodies against self because
    a. A person will not have genes for antibodies against self.
    b. Cells with TCRs against self are destroyed immediately after birth.
    c. Cells with TCRs against self are destroyed during B cell development.
    d. Cells that bind MHCs are destoyed by maternal antibodies.
    e. None of the above.

18. Which one of the following statements is **not** true?
    a. Immunological tolerance can be induced by low doses of antigen.
    b. Anergy can occur if a vaccine contains only adjuvant and not antigen.
    c. Perforin is secreted by killer T cells to lyse target cells.
    d. Perforin is secreted by natural killer cells to lyse target cells.
    e. None of the above

19. The secondary immune response is due to
    a. Memory B cells.
    b. Memory T cells.
    c. Antigen-presenting cells.
    d. Vaccination.
    e. a and b

20. Activation of B cells requires all of the following. Which is the third step?
    a  A bound antigen is taken in by the B cell
    b. A helper T cell binds an antigen-MHC II complex
    c. An antigen is digested
    d  IL-2, IL-4, IL-5, and IL-6 are released
    e. A B cell divides

21. Interleukins cause proliferation of all of the following **except**
    a. Helper T cells.
    b. Pre-killer cells.
    c. B cells.
    d. Antigen-presenting cells.
    e. None of the above.

## PROBLEMS

1. Penicillin is not antigenic. Explain why some people develop antibodies and have allergic reactions to penicillin.

2. In thrombocytopenic purpura, a drug such as aspirin coats platelets and the platelets are destroyed by complement. Provide a mechanism of action for this disease.

3. Bone marrow transplants can reject the *host* in graft-versus-host disease. Explain how this occurs. Why are bone marrow transplants done?

# Chapter 16

4. The following antibody titers against a certain disease were obtained from three patients:

   | Patient | Antibody titer | | | | | |
   |---|---|---|---|---|---|---|
   | | Week 1 | | Week 2 | | Week 3 | |
   | | IgG | IgM | IgG | IgM | IgG | IgM |
   | A | 0 | 0 | 0 | 0 | 0 | 0 |
   | B | 256 | 0 | 256 | 0 | 256 | 0 |
   | C | 0 | 512 | 0 | 1024 | 32 | 2048 |

   Which patient just got infected? Which patient has never had this disease? Explain the other patient.

5. Hepadnaviruses (e.g., Hepatitis B virus) is a DNA-containing virus that contains reverse transcriptase. What does reverse transcriptase do? Why aren't the hepadnaviruses classified as retroviruses? Name one retrovirus.

6. A patient with dysgammaglobulinemia suffers from recurrent mucus-membrane infections because they are lacking one type of antibody. Which type? Propose a mechanism whereby suppressor T cells are responsible for this disease.

7. The ratio of CD4:CD8 is 2.0 in normal individuals. What can you conclude if a patient has a CD4:CD8 ratio of 0.5?

## ANSWERS TO STUDY QUESTIONS

1. b    2. e    3. a    4. c
5. d    6. c    7. c    8. c
9. a    10. b   11. c   12. d
13. a   14. b.  15. a   16. a
17. c   18. b   19. e   20. b
21. d

# CHAPTER 17

# Neurons and Electrical Signaling

## LEARNING OBJECTIVES

Be able to
1. Describe the structure of a neuron, myelin sheath, and synapse.
2. Define axonal transport and differentiate between slow axonal transport and fast axonal transport.
3. Describe the events of an action potential including depolarization and repolarization.
4. Describe the saltatory conduction of a nerve impulse.
5. Differentiate between a chemical synapse and an electrical synapse.
6. Describe the release of a neurotransmitter.
7. Differentiate between the following:
   a. Fast chemical transmission and slow chemical transmission
   b. Excitatory and inhibitory neurotransmitters
   c. Nicotinic acetylcholine receptor and muscarinic acetylcholine receptor
8. Describe how $K^+$ channels and long-term potentiation are involved in learning.
9. Describe how opiate receptors function.
10. Describe how nitric oxide acts as a neurotransmitter.
11. Compare the mechanisms of action of photoreceptors, odor receptors, salt taste buds, and mechanoreceptors.
12. Define neurite, growth cone, and nerve growth factor.

## CHAPTER OVERVIEW

I. NEURON STRUCTURE, pp. 748-752

   A. The Discovery of Neurons Was Hindered by Their Unusual Morphology, p. 748
      1. In 1873, Golgi was the first to see nerve cells; he believed nerve cells were continuous with one another.
      2. Cajal identified the single nerve cell called a neuron.

   B. Axons and Dendrites Are Cytoplasmic Extensions of the Cell Body That Send and Receive Electrical Signals, pp. 748-749
      1. A neuron consists of a cell body and dendrites and axons.
      2. The cell body contains the nucleus, Nissl substance (ribosomes and ER), an extensive Golgi complex, mitochondria, and lysosomes.
      3. Dendrites and axons project from the cell body.
      4. Dendrites receive signals from other cells; they are highly branched, and <1 mm long.
      5. A single axon emerges from the cell body at the axon hillock; it can measure ≥1 m.
      6. The cytoskeleton consists of
         a. Microtubules in the axons with their plus ends away from the cell body; microtubules in dendrites are oriented in both directions,
         b. Neurofilaments (intermediate filaments), and
         c. Actin filaments.

## C. Glial Cells Play a Supporting Role and Produce Myelin Sheaths, pp. 749-751
1. Glia fills spaces between neuron cell bodies and around axons and dendrites.
2. Glial cells include astrocytes, oligodendrocytes, and Schwann cells.
3. Oligodendrocytes produce sheaths around the axons.
4. Virchow observed that the vertebrate central nervous system contains white matter (axons covered by myelin).
5. Oligodendrocytes wrap their plasma membranes around axons to produce myelin.
6. Schwann cells form myelin in the peripheral nervous system.
7. The myelin sheath is periodically interrupted by nodes of Ranvier.

## D. Axonal Transport Moves Materials Back and Forth Along the Axon, pp. 751-752
1. Slow axonal transport moves proteins and cytoskeletal filaments 1-5 mm/day toward the axon tip.
2. Fast axonal transport moves organelles 100-500 mm/day toward or away from the axon tip. Fast transport is mediated by kinesin and dynein.

## II. TRANSMISSION OF NERVE IMPULSES, pp. 752-768
1. Read the discussion of **action potential** on pp. 199-202.
2. Read about the **Nernst equation** on pp. 196-197.
3. An action potential is caused by a stimulus that triggers an increase in membrane permeability to $Na^+$.
4. Depolarization causes a few voltage-gated channels to open which causes further depolarization.
5. Voltage-gated $K^+$ channels open to reestablish the membrane potential.

### A. Axon Diameter and Myelination Influence the Rate at Which Action Potentials Are Propagated, pp. 753-755
1. To initiate an action potential, a stimulus must depolarize a site on the neuron plasma membrane to the threshold potential. An increase in stimulus has no further effect.
2. In a subthreshold stimulus, the $Na^+$ influx < normal $K^+$ efflux.
3. A local flow of electric current alters the membrane potential in the area around the site of the threshold stimulus.
4. Cations diffuse to negatively charged regions of the membrane creating the current that depolarizes the surrounding membrane to trigger new action potentials.
5. Action potentials move faster along large diameter axons.
6. The myelin sheath reduces ion leakage to increase the speed of new action potential formation.
7. Action potentials can only occur where the plasma membrane is exposed to extracellular fluid ($Na^+$); i.e., at the nodes of Ranvier.
8. Action potentials jump from node to node causing saltatory conduction.
9. Speed of transmission increases with increasing node distance.
10. In multiple sclerosis, myelin is destroyed in the brain.
11. In short axons (<1 mm), a nerve impulse is conducted by a current flow rather than action potentials.

### B. Neurons Communicate with Each Other at Specialized Junctions Called Synapses, pp. 755-756
1. Synapses transmit signals from the presynaptic cell to the postsynaptic cell.

# Chapter 17

2. Neurons are the presynaptic cell; postsynaptic cells include dendrites, cell bodies, and axons of other neurons, secretory cells, and muscle cells.
3. Chemical synapses release chemical neurotransmitters that diffuse across the synaptic cleft to the postsynaptic cell causing excitatory postsynaptic potentials or inhibitory postsynaptic potentials.
4. Neurotransmitters are found in synaptic vesicles and secretory vesicles in the axon terminal.
5. Electrical synapses join the pre- and postsynaptic cells by gap junctions.
6. Membrane depolarization passes directly through the gap junctions.

C. **Neurons Employ a Variety of Different Chemicals as Neurotransmitters, p. 757**
   1. Neurotransmitters are released from nerve endings upon stimulation and evoke the same response as a normal nerve impulse in a postsynaptic cell.

D. **Neurotransmitters Are Stored in Vesicles That Discharge Their Contents into the Synaptic Cleft, pp. 757-759**
   1. Release of a few vesicles causes a miniature postsynaptic potential.
   2. Release of hundreds of vesicles causes an excitatory postsynaptic potential.
   3. Excessive stimulation depletes the axon's supply of vesicles.

E. **Neurotransmitter Release Is Regulated by the Entry of Calcium Ions into the Axon Terminal, pp. 759-760**
   1. Membrane depolarization causes voltage-gated $Ca^{2+}$ channels in the presynaptic plasma membrane to open.
   2. $Ca^{2+}$ diffuses into the cytosol triggering exocytosis of vesicles.
   3. A releasable pool of vesicles is bound to the inner surface of the plasma membrane by a channel-forming protein.
   4. $Ca^{2+}$ causes the channel to open to release the vesicles.
   5. A reserve pool of vesicles is attached to the cytoskeleton by synapsin.
   6. Reserve vesicles are released when a $Ca^{2+}$-calmodulin-dependent protein kinase phosphorylates synapsin.

F. **Synaptic Vesicles Are Recycled after Fusing with the Presynaptic Membrane, pp. 760-761**
   1. In exocytosis, vesicles fuse with the plasma membrane to release their contents.
   2. Clathrin-coated vesicles form from the plasma membrane. The coated vesicles form new synaptic vesicles.

G. **Neurotransmitters Bind to Postsynaptic Receptors That Mediate Either Fast or Slow Chemical Transmission, p. 761**
   1. Excitatory neurotransmitters trigger depolarization.
   2. Inhibitory neurotransmitters cause hyperpolarization.

|  |  | Fast transmission | Slow transmission |
|---|---|---|---|
| Neurotransmitters |  |  |  |
|  | Excitatory | Acetylcholine | Amines |
|  |  | Glutamate | Peptides |
|  | Inhibitory | Glycine, GABA |  |
| Speed |  | 10ths of msec. | 100s of msec. |
| Mechanism |  | Gated-ion channels | G proteins |

# Chapter 17

**H. Acetylcholine Is Involved in Both Fast Excitatory Transmission and Slow Inhibitory Transmission, pp. 761-762**

|  | Skeletal muscle | Heart muscle |
|---|---|---|
| Neurotransmitters | Acetylcholine (Ach) | Acetylcholine (Ach) |
| Mechanism | Ion channel | G protein activated |
| Receptor | Nicotinic Ach receptor | Muscarinic Ach receptor |
| Speed | Fast | Slow |

**I. GABA and Glycine Mediate Fast Inhibitory Transmission, pp. 762-763**
1. GABA and glycine open Cl⁻ channels.
2. Cl⁻ enters the cell to cause hyperpolarization.
3. In presynaptic inhibition, inhibitory neurons terminate on the axons of excitatory neurons.

**J. Several Neurotransmitters Bind to Receptors That Influence Adenylyl Cyclase, p. 764**
1. Dopamine elevates cAMP in postsynaptic cells.
2. Parkinson's disease is associated with degeneration of dopamine-releasing neurons.
3. Certain types of schizophrenia are due to excess dopamine. (Thorazine blocks dopamine receptors.)

**K. Cyclic AMP-Dependent Phosphorylation of a Potassium Channel Is Associated with a Single Type of Learning, pp. 764-765**
1. The sea slug *Aplysia* can become habituated and therefore not respond to a stimulus or sensitized so it responds to a stimulus.
2. Sensitization is a simple type of learning due to facilitator neurons.
3. Facilitator neurons release serotonin to a sensory neuron.
   a. Serotonin activates adenylyl cyclase and cAMP.
   b. cAMP activates protein kinase A which phosphorylates (and closes) a $K^+$ channel.
   c. The action potential is prolonged and more neurotransmitter is released.
   d. The sensory neuron's signal becomes stronger with time.

**L. Enkephalins and Endorphins Inhibit Pain-Signaling Neurons by Binding to Opiate Receptors, pp. 765-766**
1. Binding of morphine, enkephalins, or endorphins to opiate receptors in the plasma membrane of the central nervous system block pain.
2. Two 5-amino-acid peptides (i.e., met-enkephalin and leu-enkephalin) and larger peptides (e.g., endorphins) are produced by nerve tissue and bind to opiate receptors.
3. Opiate receptors are G protein-linked receptors.
4. Enkephalin binding opens $K^+$ channels so the membrane hyperpolarizes and closes $Ca^{2+}$ channels to inhibit neurotransmitter release.

**M. Nitric Oxide Is a Novel Type of Neurotransmitter That Acts by Stimulating Cyclic GMP Formation, pp. 766-768**
1. Arginine $\xrightarrow{\text{NO synthase}}$ nitric oxide + citrulline
2. Ach binds to endothelial cells lining blood vessels to relax adjacent muscle cells and dilate blood vessels:
   a. Ach binds G protein-linked receptors

b. IP$_3$ triggers release of Ca$^{2+}$ from the ER
c. Ca$^{2+}$-calmodulin stimulates NO synthase
d. Nitric oxide (NO) gas diffuses to smooth muscle cells
e. NO activates guanylyl cyclase to form cGMP
f. cGMP activates protein kinase G which phosphorylates muscle proteins.
3. NO may be involved in long-term potentiation (LTP) which is a means of storing memories.
a. LTP involves repeated transmission across a synapse.

### III. DETECTING STIMULI AND TRIGGERING RESPONSES, pp. 768-775

**A. Vertebrate Photoreceptors Detect Light Using a Mechanism That Involves Cyclic GMP, pp. 768-769**
1. Sensory cells perceive changes in the environment:
   a. Photoreceptors (rods and cones) are sensitive to light
   b. Mechanoreceptors respond to pressure or movement
   c. Chemoreceptors detect chemicals
   d. Thermoreceptors monitor temperature
2. Photoreceptors contain rhodopsin (11-cis-retinal bound to opsin).
3. Rhodopsin $\xrightarrow{photon}$ Altered rhodopsin → G protein → breakdown of cGMP → Na$^+$ channels close → hyperpolarization → perception of light.

**B. Sensory Stimuli Alter the Membrane Potential of Sensory Cells by a Variety of Different Mechanisms, p. 769**
1. Odor molecules bind G$_{olf}$ to produce cAMP and membrane depolarization.
2. Na$^+$ in foods diffuses through Na$^+$ channels causing depolarization directly.
3. Movement causes stereocilia in the inner ear to open cation channels and trigger depolarization.

**C. Responses Triggered by the Nervous System Include Muscular Contraction, Glandular Secretion, and Neurosecretion, pp. 769-771**
1. In the heart, Ach decreases membrane excitability to slow the heart rate and norepinephrine increases membrane excitability to speed the heart.
2. Epinephrine causes increased heart rate, blood pressure, breathing rate, and blood glucose concentration.
3. Some neurons secrete hormones into the blood by neurosecretion.
a. Membrane depolarization triggers exocytosis of hormones for neurosecretion.

### IV. NEURON GROWTH AND DEVELOPMENT, pp. 771-775

**A. The Growth Cone Directs the Outgrowth of Neurites, pp. 771-772**
1. During embryonic development, some ectodermal cells develop into neuroblasts.
2. Neuroblasts stop dividing and develop cytoplasmic extensions called neurites.
3. Neurites will become axons and dendrites.
4. The tip of each axon has a growth cone.
5. The growth cone has filopodia which adhere to new surfaces and cause the axon to grow.

- B. **Neurite Growth Is Stimulated by Nerve Growth Factor as well as a Variety of Other Proteins, p. 772**
    1. Nerve growth factor (NGF) promotes neurite outgrowth, especially in sensory neurons and sympathetic neurons.
    2. NGF activates a protein-tyrosine kinase.

- C. **Axons Are Guided to Their Proper Destination by Cell-Cell Contacts, Matrix Molecules, and Diffusible Substances, pp. 772-774**
    1. Neurons have surface molecules that guide them to their targets (e.g., cell body, dendrite, axon) to form synapses.
    2. Filopodia change direction when they make contact with guidepost cells.
    3. Growth cones use fasciclins on neighboring axons as scaffolds upon which to extend.

- D. **Most Mature Neurons Lose the Capacity to Divide, pp. 774-775**
    1. Damaged or cut neurons can grow a new axon.
    2. Neuroblast tumors (neuroblastomas) divide in culture and can be induced to make dendrites, axons, and neurotransmitters.

# KEY TERMS

| | | |
|---|---|---|
| actin filament | action potential | astrocyte |
| axon | axon terminal | cell body |
| chemical synapse | chemoreceptor | cone |
| dendrite | dopamine | electrical synapse |
| enkephalins | excitatory postsynaptic potential | facilitator neuron |
| fasciclin | fast axonal transport | fast chemical transmission |
| G protein | glia | growth cone |
| long-term potentiation | mechanoreceptor | microtubule |
| miniature postsynaptic potential | muscarinic acetylcholine receptor | nerve growth factor |
| nerve impulse | neurite | neuroblastoma |
| neurofilament | neurosecretion | neuron |
| neurotransmitter | nicotinic acetylcholine receptor | nitric oxide |
| nodes of Ranvier | oligodendrocyte | opiate receptor |
| photoreceptor | receptor | rhodopsin |
| rod | Schwann cell | secretory vesicle |
| sensory cell | slow axonal transport | slow chemical transmission |
| synapse | synapsin | synaptic cleft |
| synaptic vesicle | thermoreceptor | threshold potential |

# Chapter 17

## KEY FIGURE

Label the axon, synaptic cleft, gap junctions, synaptic vesicles, receptors, and postsynaptic cells. Which figure shows chemical transmission? Electrical transmission?

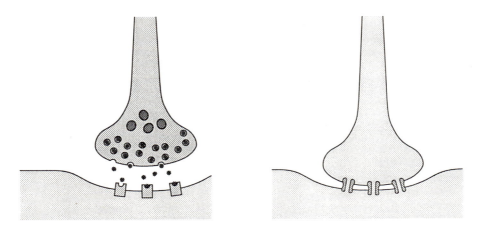

## STUDY QUESTIONS

1. The cell body contains all of the following **except**
   a. Ribosomes.
   b. ER.
   c. Golgi complex.
   d. Mitochondria.
   e. None of the above.

2. Which one of the following is **not** true?
   a. The plus ends of microtubules are pointed toward the cell body in axons.
   b. The plus ends of microtubules are pointed away from the cell body in axons.
   c. The plus ends of microtubules are always pointed toward the cell body in dendrites.
   d. The plus ends of microtubules are always pointed away from the cell body in dendrites.
   e. None of the above.

3. Release of acetylcholine causes
   a. $Na^+$ efflux.
   b. $Na^+$ influx.
   c. $K^+$ efflux.
   d. $K^+$ influx.
   e. None of the above.

4. Synaptic vesicles form from
   a. Microtubules in the axon.
   b. Coated pits in the plasma membrane.
   c. Fusion of lysosomes.
   d. Accumulations of neurotransmitter.
   e. None of the above.

Use the following diagram to answer questions 5-8.

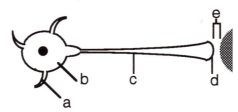

5. Which letter indicates where neurotransmitters are released?

6. Microtubules move materials 100 mm/day in this structure.

7. The myelin sheath will be found here?

8. The gray-shaped cell could be all of the following **except**
   a. A muscle cell.
   b. A cell body.
   c. A dendrite.
   d. An epithelial cell.
   e. A secretory cell.

9. Which of the following conditions occur in depolarization?
   a. $Na^+$ channels closed—$Cl^-$ channels open
   b. $Na^+$ channels closed—$K^+$ channels closed
   c. $Na^+$ channels closed—$K^+$ channels opened
   d. $Na^+$ channels opened—$K^+$ channels closed
   e. None of the above

10. Which graph shows the effect of a subthreshold excitatory stimulus? Assume the neuron has a resting potential of -60 mV and a threshold potential of -30 mV.

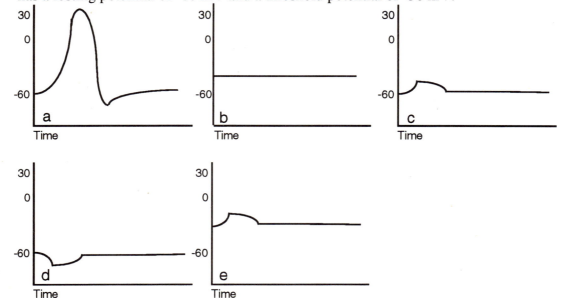

11. Serotonin causes
    a. Release of neurotransmitter.
    b. Blocking of pain.
    c. Synthesis of cAMP.
    d. Production of acetylcholinesterase.
    e. None of the above.

12. The speed of an action potential is proportional to all of the following **except**
    a. Size of stimulus.
    b. Internodal distance.
    c. Myelination.
    d. Axonal diameter.
    e. None of the above.

13. Fast inhibitory neurotransmitters cause
    a. $Cl^-$ efflux.
    b. $Cl^-$ influx.
    c. $K^+$ efflux.
    d. $Na^+$ efflux.
    e. $Na^+$ influx.

14. The polarized state of a resting neuron is due to
    a. $Ca^{2+}$.
    b. $Cl^-$.
    c. $K^+$.
    d. $Na^+$.
    e. None of the above.

Use the following choices to answer questions 15-17.
    a. The heart would beat faster.
    b. The heart would stop.
    c. Skeletal muscle spasms.
    d. Flaccid paralysis of skeletal muscles.
    e. b and d.

15. Cobratoxin blocks nicotinic acetylcholine receptors. What effect will this have on its prey?

16. The cold-drug atropine blocks muscarinic acetylcholine receptors. You would expect this drug to cause _____.

17. The poisonous fungus *Amanita muscaria* produces muscarine. You would expect the symptoms of this mushroom poisoning to be _____.

18. The drug PCP binds glutamate receptors to cause
    a. Activation of G protein.
    b. $IP_3$ production.
    c. $Ca^{2+}$ influx.
    d. Breakdown of cGMP.
    e. None of the above.

19. In an unstimulated nerve,
    a. The inside is negatively charged compared to the outside.
    b. The outside is negatively charged compared to the inside.
    c. The intracellular and extracellular charges are equal.
    d. It depends on the nerve.

20. Saltatory conduction occurs in
    a. Myelinated nerves.
    b. Unmyelinated nerves.
    c. Gray matter.
    d. Small diameter nerves only.
    e. None of the above.

21. Microscopic examination of axons reveals vesicles aligned in rows through the axon from the cell body to the axon terminal. Which one of the following do you predict?
    a. Cytochalasin will prevent movement of the vesicles.
    b. Secretory vesicles are being moved to the axon terminal.
    c. Slow axonal transport is moving the vesicles.
    d. The vesicles contain digestive enzymes.
    e. None of the above

# PROBLEMS

1. Use the Nernst equation (pp. 196-197) to calculate the membrane potential of an axon:

    |  | $[K^+]$ (mM) |
    |---|---|
    | Inside the cell | 140 |
    | Outside the cell | 5 |

    What is the effect of adding more $K^+$ outside the cell?

2. Compare and contrast the effects of the two *Clostridium* bacteria toxins: botulinum toxin and tetanus toxin. Botulinum toxin blocks the release of acetylcholine and tetanus toxin blocks glycine receptors.

3. What conclusion can you draw from the following data?

    | Experiment | Observations | |
    |---|---|---|
    | #1 | Retinal growth cones cross other retinal neurites without hesitation. | Retinal growth cones retract when they encounter a motor neuron. The neurite then grows a new growth cone. |
    | #2 | The filopodia of one growth cone contact 25 different axons. | Invariably the growth cone follows one axon. |
    | #3 | Growth cones secrete proteases. | Glial cells inhibit proteases. |
    | #4 | The growth cone finds its correct target even when the axon is severed near the cell body. | |

## Chapter 17

4. Organophosphate insecticides are competitive inhibitors of acetylcholinesterase. What is the effect of organophosphates on an animal? Which of the following drugs could be used to counteract organophosphate poisoning?

   | Drug | Method of action |
   |---|---|
   | Atropine | Blocks muscarinic acetylcholine receptors |
   | Curare | Competitive inhibition of acetylcholine receptors |
   | Physostigmine | Competitive inhibition of acetylcholinesterase |

5. The toxins of poison arrow frogs are listed below. Complete the table to show the symptoms of poisoning by each toxin.

   | Toxin | Molecular action | Effect on neuron |
   |---|---|---|
   | Batrachotoxin | Prevents $Na^+$ gate closure | |
   | Pumilotoxin B | Prevents reuptake of $Ca^{2+}$ | |
   | Histrionicotoxin | Prevents $K^+$ efflux | |

6. The sea slug *Hermissenda* is exposed to a flash of light and shaking aquarium. The sea slug uses its muscular foot to attach itself firmly as it would when an ocean wave hits. After 150 training cycles, the slug attaches itself when the light flashes without the accompanying movement. Why does the slug respond to light in this way since it doesn't present the same threat as an ocean wave? Injections of protein kinase C cause the same "foot reaction." What can you conclude from this? What do you expect to happen if protein kinase C is blocked in the slug?

7. Changing the [$Na^+$] outside the cell doesn't change the resting membrane potential. Why not?

## ANSWERS TO STUDY QUESTIONS

1. e    2. a    3. b    4. b
5. d    6. c    7. c    8. d
9. d    10. c   11. c   12. a
13. b   14. c   15. d   16. b
17. b   18. c   19. a   20. a
21. b

# CHAPTER 18

## The Cancer Cell

### ⇧ LEARNING OBJECTIVES

Be able to
1. Define the following terms:
    a. Cancer
    b. Tumor
    c. Proto-oncogene
    d. Oncogene
2. Differentiate between a benign and a malignant tumor.
3. Describe the process of metastasis.
4. Describe the immune response to cancer cells.
5. List the differences between normal and cancer cells.
6. Describe how oncogenes are activated by the following mechanisms:
    a. Oncogenic DNA viruses
    b. Oncogenic RNA viruses
    c. Point mutations
    d. DNA rearrangement
    e. Insertional mutagenesis
    f. Gene amplification
    g. Chromosomal translation
    h. Radiation
    i. Chemical carcinogens
7. Describe how each of the following are involved in malignancies:
    a. Growth factors
    b. Protein-tyrosine kinases
    c. G proteins
    d. Protein-serine/threonine kinases
    e. Transcription factors
8. Define tumor suppressor genes and provide an example.
9. Explain the Ames Test.
10. Describe how epidemiological data can be used to prevent cancer.
11. Describe the role of the following in preventing or curing cancer:
    a. Antioxidants
    b. Surgery
    c. Radiation
    d. Chemotherapy
    e. Immunotherapy

### ⇧ CHAPTER OVERVIEW

I. **WHAT IS CANCER? pp. 778-788**
   1. Cancer describes a disease in which tissues grow and spread unrestrained; the term refers to malignant tumors.
   2. Carcinomas arise from epithelial cells.
   3. Sarcomas arise in supporting tissues.
   4. Lymphomas and leukemias arise from blood and lymphatic cells.

   A. **Tumors Arise When the Rate of Cell Division Exceeds the Rate of Cell Differentiation and Loss, p. 778**
      1. A growing tissue mass is called a tumor or neoplasm.

# Chapter 18

2. In normal cells, the rate of cell division is balanced with the rate of cell differentiation and loss.
3. In cancer cells, the rate of cell division exceeds the rate of cell loss.

B. **Malignant Tumors Are Capable of Invasion and Metastasis, pp. 778-781**
1. Benign tumors grow in a confined local area and are rarely life-threatening.
2. Malignant tumors can invade surrounding tissues and enter the circulatory system allowing them to spread by metastasis.
3. Tumors are named by adding the suffix "–oma" to the cell type.
4. Metastasis is a multistep process:
    a. Cancer cells invade surrounding tissues and vessels: cell-cell adhesiveness is diminished in cancer cells, cancer cells are mobile, and cancer cells use proteases to digest the extracellular matrix.
    b. Cancer cells are transported to distant sites by the circulatory system.
    c. Cancer cells selectively reinvade and grow at particular sites.
5. The initial tumor cell population is heterogeneous; the cells can metastasize to a variety of tissues.

C. **The Immune System Can Inhibit the Process of Metastasis, pp. 781-782**
1. Killer T cells are produced against the altered MHCs on tumor cells. However, tumor cells may carry MHCs not recognized by the immune system.

D. **A Group of Characteristic Traits Are Shared by Cancer Cells, pp. 782-788**
1. **Cancer cells have a distinctive appearance, pp. 782-783**
    a. Malignant cells lose differentiation and proper orientation (anaplasia).
    b. Malignant cervical cells have irregular nuclei and altered size and shape, as seen in a PAP smear.
2. **Cancer cells produce tumors when injected into laboratory animals, pp. 783-784**
   Human cancer cells are injected into nude mice (lacking the thymus gland).
3. **Cancer cells are immortal in culture, p. 784**
4. **Cancer cells grow to high densities in culture, pp. 784-785**
    a. Normal cells stop growing and moving when they produce a monolayer.
    b. Cancer cells are less susceptible to contact inhibition or density-dependent inhibition of growth.
    c. Some cancer cells produce their own growth factors such as transforming growth factors.
5. **Cancer cell growth is anchorage independent, pp. 785-786**
   Unlike normal cells, cancer cells grow suspended in a liquid medium.
6. **Cancer cells lack normal cell cycle controls, p. 786**
   When normal cells stop growing due to lack of nutrients or growth factors, they stop near the end of $G_1$, in $G_0$ state. Cancer cells keep growing in suboptimal conditions and die at random points in the cell cycle.
7. **Cancer cells exhibit cell-surface alterations, pp. 786-787**
    a. Cancer cells lack fibronectin and do not stick to one another.
    b. Cancer cells agglutinate with lectins because their lectin receptors are more mobile than in normal cells.
    c. Cancer cells have fewer gap junctions.

# Chapter 18

      d. Cancer cells have tumor-specific cell-surface antigens. During immune surveillance, the immune system destroys malignant cells.
- 8. **Cancer cells secrete proteases, embryonic proteins, and proteins that stimulate angiogenesis, pp. 787-788**
  - a. Cancer cells secrete plasminogen activator which produces the protease plasmin.
  - b. Cancer cells produce the embryonic proteins α-fetoprotein, carcinoembryonic antigen, chorionic gonadotropin and placental lactogen.
  - c. Cancer cells stimulate the formation of blood vessels (angiogenesis).
- 9. **Cancer cells exhibit chromosomal abnormalities, p. 788**
  - a. The Philadelphia chromosome (due to a translocation of a piece of chromosome 22 to chromosome 9) occurs in 90% of chronic granulocytic leukemias.

## II. WHAT CAUSES CANCER? pp. 788-802

### A. Chemicals Can Cause Cancer, pp. 788-789
1. Cancer-causing chemicals are called carcinogens.

### B. Chemical Carcinogenesis Involves Initiation and Promotion Stages, pp. 789-791
1. In the 1940s, Rous observed that initiation causes irreversible conversion to the preneoplastic state and promotion stimulates the preneoplastic cells to divide and form tumors.
2. Initiation appears to involve mutations.
3. Promotion requires prolonged exposure to and the presence of promoting agents. Promotion is reversible.

### C. Radiation Can Cause Cancer, pp. 791-792
1. Ultraviolet radiation is absorbed by normal skin pigmentation; it does not pass through the skin.
2. X-rays and emissions from radioactive elements penetrate the human body.
3. Radiant energy initiates cancer by causing mutations.

### D. Viruses Cause a Variety of Different Animal Cancers, p. 792
1. Oncogenic viruses cause some cancers.
2. Oncogenic viruses enter a latent period when their nucleic acid is present in the cell but not reproduced.
3. Cancer occurs after exposure to radiation, chemical carcinogens, hormones, or other viruses.

### E. Viruses Have Been Implicated in Only a Few Types of Human Cancer, pp. 792-793

| Human cancer | Virus | DNA or RNA? |
| --- | --- | --- |
| Burkitt's lymphoma | EB virus | DNA |
| Nasopharyngeal cancer | EB virus | DNA |
| Liver cancer | Hepatitis B virus | DNA |
| Uterine cervical cancer | Human Papilloma virus | DNA |
| Acute T-cell leukemia | Human T-cell leukemia virus | RNA |

# Chapter 18

**F. Oncogenic Viruses Insert Their Genetic Information into the Chromosomal DNA of Infected Cells, pp. 793-794**
  1. The DNA of oncogenic DNA viruses usually becomes incorporated into the host cell's chromosome.
  2. Oncogenic RNA viruses use reverse transcriptase to synthesize double-stranded viral DNA which can become incorporated into the host cell's chromosome.

**G. Viral Oncogenes Are Altered Versions of Normal Cellular Genes, pp. 794-795**
  1. Genes that induce malignant transformations are called oncogenes.
  2. Normal cellular genes that resemble oncogenes are called proto-oncogenes.

**H. Cellular Oncogenes Arise from the Mutation of Proto-Oncogenes, pp. 795-797**

| Mechanism | Example | Result | Human disease |
|---|---|---|---|
| Point mutation | H-*ras* | Abnormal Ras protein differs in one amino acid | Lung carcinoma |
| DNA rearrangement | *trk* | A fusion or hybrid protein forms | Thyroid carcinoma |
| Insertional mutagenesis | c-*erbB* | Insertion of provirus converts gene | Chicken erythroleukemia |
| Gene amplification | K-*sam* | Increased number of gene copies | Stomach carcinoma |
| Chromosomal translocation | c-*myc* | Movement of a piece of chromosome 8 to chromosome 14 | Burkitt's lymphoma |

**I. Oncogenes Code for Proteins Involved in Normal Growth Control, pp. 797-799**
  1. Some oncogenes code for growth factors which result in autocrine stimulation.
  2. Some oncogenes code for abnormal (i.e., unregulated) growth factor receptors. Other oncogenes code for nonreceptor protein-tyrosine kinases. These kinases excessively phosphorylate cellular proteins.
  3. Some oncogenes code for abnormal membrane-associated G proteins. **Read about G proteins on pp. 211-224.**
  4. Protein-serine/threonine kinases are encoded by some oncogenes.
  5. Transcription factors are encoded by some oncogenes.

**J. Oncogenes Act at Several Points in the Same Growth Control Pathway, p. 799**
  1. Oncogenes disrupt normal growth control by producing abnormal versions or abnormal quantities of proteins.

**K. Cancer Can be Induced by the Loss of Tumor Suppressor Genes that Normally Inhibit Cell Proliferation, pp. 799-801**
  1. A normal tumor suppressor gene restrains cell growth and division; this function is lost in abnormal tumor suppressor genes.

# Chapter 18

    2. *p53* is a tumor suppressor gene that is altered in about half of all human malignancies.
    3. p53 protein is a nuclear transcription factor that causes $G_1$ arrest.

**L. Human Cancers Develop by the Stepwise Accumulation of Mutations Involving Both Oncogenes and Tumor Suppressor Genes, p. 801**

**M. An Increased Risk of Developing Cancer Can Be Inherited, pp. 801-802**

**N. Malignancy Is Not Always an Irreversible Change, p. 802**
    1. In crown gall disease, the Ti plasmid induces excessive cell growth. Crown gall cells grow normally when grafted onto healthy plants and abnormally when grown in cell culture.
    2. Cancer cells from mouse teratocarcinoma grow normally when injected into a normal blastula.

## III. CAN CANCER BE PREVENTED OR CURED? pp. 802-809

**A. Epidemiological Data Allow Potential Carcinogens To Be Identified in Exposed Human Populations, pp. 803-804**

**B. The Ames Test Is a Rapid Screening Method for Identifying Potential Carcinogens, pp. 804-805**
    1. The Ames Test is used to determine the mutagenic potential of a chemical. It is based on the assumption that chemicals that are not mutagenic will not be carcinogenic.
    2. The his⁻ bacteria used in the test are only able to grow if histidine is provided in the growth medium or if they undergo a reversion from his⁻ to his⁺. The amount of bacterial growth seen is an indication of the frequency of mutations produced.
    3. To assay the effects of human metabolism on chemicals, chemicals can first be mixed with liver enzymes before being screened in the Ames Test.

**C. Minimizing Exposure to Carcinogens Requires Changes in Lifestyle, pp. 805-806**

**D. Dietary Antioxidants May Help to Counteract the Effects of Some Carcinogens, pp. 806-807**
    1. Many carcinogens cause the production of free radicals. Antioxidants will donate electrons to free radicals so they cannot harm cells.

**E. Surgery, Radiation, and Chemotherapy Are the Conventional Treatments for Cancer, pp. 807-808**
    1. Radiation therapy and chemotherapy work by causing mutations. Actively growing cells, e.g., cancer cells, will be more susceptible to these mutations than normal cells in the $G_0$ state.
    2. Chemotherapeutic agents must affect eukaryotic cells and will have effects on all growing cells in the body; e.g., cancer cells, intestinal mucosa, hair follicles, blood cells.

**F. Immunotherapy Is an Experimental Treatment Designed To Selectively Destroy Cancer Cells, pp. 808-809**
    1. A person's own lymphocytes can kill cancer cells.

# Chapter 18

2. These include lymphokine-activated killer cells and tumor-infiltrating lymphocytes.

## KEY TERMS

anaplasia
benign
crown gall disease
immune surveillance theory
malignant
nonreceptor protein-tyrosine kinase
oncogenic
phorbol esters
proto-oncogene
retrovirus
teratocarcinoma

angiogenesis
carcinogen
free radical
leukemia
metastasis
nude mice
p53
protein-serine/threonine kinase
radiation
reverse transcriptase
tumor

antioxidant
carcinoma
G protein
lymphoma
neoplasm
oncogene
PAP smear
protein-tyrosine kinase
receptor protein-tyrosine kinase
sarcoma
tumor suppressor genes

## KEY FIGURE

Label the nucleic acids and steps to show the replication of a retrovirus.

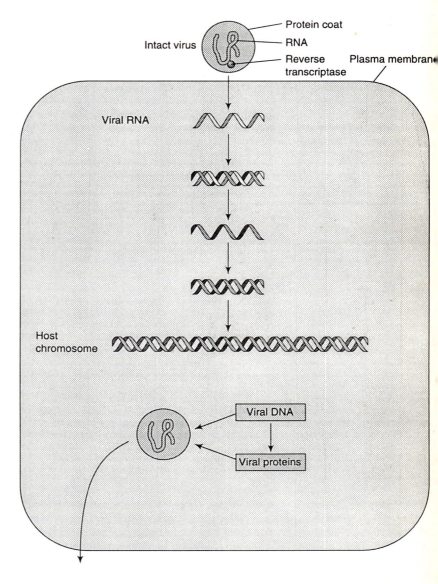

# Chapter 18

## 🏠 STUDY QUESTIONS

1. Which one of the following is **not** correctly matched?
   a. Carcinoma — ~~blood cells~~ bone cells
   b. Leukemia — blood cells
   c. Lymphoma — lymphatic cells
   d. Sarcoma — connective tissue
   e. Sarcoma — ~~supporting~~ tissue
      muscle

2. Which one of the following **leads** to the others?
   a. Cells aren't susceptible to density-dependent inhibition.
   b. Cells do not enter the $G_0$ state.
   c. Cells do not recognize contact inhibition.
   d. Cells grow in low nutrient concentrations.
   e. None of the above.

3. Which of the following is **not** a characteristic of cancer cells?
   a. Lack fibronectin
   b. Produce plasminogen activator
   c. Increased lectin receptors
   d. Produce chorinic gonadotropic hormone
   e. None of the above

4. In cell culture, cancer cell lines
   a. Require the constant addition of nutrients.
   b. Require the constant addition of growth factors.
   c. Must be transferred to new growth medium after a few cell divisions.
   d. Stop growing at the monolayer stage.
   e. None of the above.

5. Addition of fibronectin to malignant cells would have all of the following effects **except** which one? The malignant cells would
   a. Flatten out.
   b. Move when in contact with other cells.
   c. Not recognize density-dependent inhibition of growth.
   d. Adhere to one another.
   e. None of the above.

6. Turpentine does not cause cancer, however it can cause tumors to appear. This indicates that turpentine is
   a. An initiator.
   b. A carcinogen.
   c. A promotor.
   d. A mutagen.
   e. None of the above.

7. All of the following cancers are caused by viruses **except**
   a. Chicken leukemia.
   b. Chicken sarcoma.
   c. Feline leukemia.
   d. Mouse mammary gland cancer.
   e. None of the above.

## Chapter 18

8. All of the following are methods of activating oncogenes **except**
   a. Chromosomal translocation.
   b. DNA rearrangement.
   c. Gene amplification.
   d. Point mutation.
   e. Stimulating of cell division.

9. All of the following are true about human RNA viruses. Which one of the following is necessary for an RNA virus to be oncogenic?
   a. Ability to replicate in a human cell
   b. Possess a protein coat
   c. Possess reverse transcriptase
   d. Possess genes
   e. All of the above

10. Oncogenic DNA viruses differ from oncogenic RNA viruses in that the RNA viruses
    a. Can remain latent in a host cell.
    b. Possess oncogenes.
    c. Possess reverse transcriptase.
    d. Can replicate in human cells.
    e. None of the above.

11. The protein-tyrosine kinase activity is 30 to 50 times greater in cancer cells than in normal cells suggesting that
    a. Cancer is due to excessive phosphorylation of proteins.
    b. Cancer is due to gene amplification.
    c. Cancer to due to growth factor production.
    d. Sarcoma cells produce src protein and normal cells do not.
    e. None of the above.

12. In cancer cells, the Ras proteins are bound to GTP more often than G proteins in normal cells. This suggests that cancer is due to
    a. Excessive amounts of G protein.
    b. Excessive amounts of GTP.
    c. G protein activation.
    d. G protein inhibition.
    e. None of the above.

13. Which one of the following acts in the nucleus to produce a malignant transformation?
    a. Growth factors
    b. Receptor protein-tyrosine kinase
    c. G proteins
    d. Protein-serine/threonine kinase
    e. Transcription factor

14. The cause of cancer is best described by which **one** of the following statements?
    a. Cancer is the result of a single mutation.
    b. Cancer is a multistep accumulation of mutations.
    c. Cancer is the result of a poor diet.
    d. Cancer is the result of a viral infection.
    e. None of the above.

15. Cells from a crown gall produce abnormal tumor cells in cell culture. When the cells are grafted onto a healthy plant, they produce normal cells. You can conclude that
    a. Cancer is caused by changes in the nucleus.
    b. The absence of promotor factors causes normal growth.
    c. The presence of normal growth factors causes normal growth.
    d. Cancer is inherited.
    e. b and c.

16. The purpose of the Ames test is to
    a. Identify carcinogens.
    b. Identify mutagens.
    c. Grow his$^+$ bacteria.
    d. Convert his$^-$ bacteria to his$^+$ bacteria.
    e. None of the above.

17. All of the following are true about chemotherapeutic drugs used against cancer **except**
    a. They are mutagenic.
    b. They affect eukaryotic cells.
    c. They are specific for viruses.
    d. They are not specific for tumors.
    e. None of the above.

18. Childhood neuroblastoma tumor cells have tumor specific transplanation antigen and no MHC I antigens. From this, you could predict that these cells
    a. Are invasive.
    b. Have an increased growth rate.
    c. Escape natural killer cells.
    d. Lack contact inhibition.
    e. None of the above.

19. Samples from three females (#1, 2, and 3) from a family with familial breast cancer were examined by Southern blotting for gene *nm23* (shown in gel A) and a mutant form of the oncogene *p53* (shown in gel B). Patients 1 and 3 do not have breast cancer; patient 2 does.

    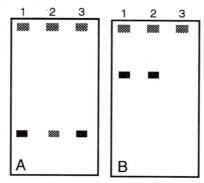

    The results indicate that
    a. *Nm23* is a tumor suppressing gene.
    b. The mutated *p53* is inherited.
    c. The mutated *p53* always causes cancer.
    d. A spontaneous mutation occurred.
    e. None of the above.

20. The following are mechanisms of action of anticancer drugs. Which one is specific for tumor cells?
    a. Blocking the mitotic spindle
    b. Hydrolyzing asparagine
    c. Inhibiting microtubule formation
    d. Inhibiting RNA synthesis
    e. None of the above

21. What is the most likely explanation for the incidence of cancer shown in this graph?

    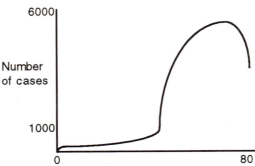

    a. Cancer is caused by a spontaneous mutation.
    b. Cancer is a multistep process.
    c. Cancer is a disease of old people.
    d. Abnormal genes are inherited.
    e. None of the above.

## PROBLEMS

1. Approximately 100,000 childbearing-aged women in the United States are infected with HIV, and an estimated 7,000 infants are born to HIV-positive mothers each year. In the United States, the rate of perinatal transmission of HIV among mothers who do not receive antiretroviral therapy is 15% to 30%. Treatment with zidovudine (ZDV) may substantially reduce the risk for perinatal HIV tranmission. Data from the Antiretroviral Pregnancy Registry are reported below:

    |  | Number |
    |---|---|
    | Pregnant women treated with ZDV | 198 |
    | Women still awaiting delivery | 30 |
    | Women for whom delivery information is available | 121 |
    | ZDV treatment during first trimester | 54 |
    | Number of infants born | 45 |
    | Therapeutic abortions | 8 |
    | Infants with birth defects | 1 |

    What is a potential risk of treating a pregnant woman with ZDV, a thymidine analog? What can you conclude from these data?

2. EB virus is known to cause Burkitt's lymphoma in regions in Africa with malaria. In the United States, EB viral infectious mononucleosis is common and Burkitt's lymphoma is rare. However, Burkitt's lymphoma occurs in AIDS patients. Explain the occurence of Burkitt's lymphoma.

# Chapter 18

3. Explain how radiation can be used to cure cancer as well as cause the disease.

4. Women with the normal *BRCA 1* gene do not inherit familial breast cancer, while women who inherit a mutated form of *BRCA 1* are predisposed to developing breast cancer. Propose a mechanism of action for this gene.

5. R. Palmer Beasley conducted a study of liver cancer in Taiwan. Between 1975 and 1986 he collected information on 22,707 male government bureaucrats. Some of his data are shown below:

|  | Subjects who tested negative for hepatitis B | Subjects with chronic hepatitis B |
|---|---|---|
|  | 19,253 | 3,454 |
| Subjects with liver cancer | 9 | 152 |

What can you conclude from these data? Can you conclude that the incidence of liver cancer is higher in males than in females? Higher in Taiwan than in another country?

6. Tumor cells lose their ability to be invasive when the gene for a cadherin protein is inserted into them. (Cadherins are calcium-ion dependent cell adhesion molecules.) Describe how gene therapy for cancer could be done using a retrovirus.

7. The following data are from an Ames Test. The amount of bacterial growth seen in the Ames Test is an indication of the amount of mutations produced. The following data show that 2-aminofluorene, a component of cigarette smoke, causes mutations. In the other experiments, 2 aminofluorene was first mixed with intestinal enzymes or the microsomal fraction of liver cells for 30 minutes before use in the Ames Test. Explain the results.

|  | Amount of bacterial growth |
|---|---|
| Control (no chemical) | 5 |
| 2-aminofluorene | 172 |
| Intestinal enzymes + 2-aminofluorene | 350 |
| Liver microsomes + 2-aminofluorene | 942 |

## ANSWERS TO STUDY QUESTIONS

1. a    2. b    3. c    4. e
5. b    6. c    7. e    8. e
9. c    10. c    11. a    12. c
13. e    14. b    15. e    16. b
17. c    18. c    19. a    20. e
21. b